R
724
C545
1993
v. 2
Porter

THE CODIFICATION OF MEDICAL MORALITY

Philosophy and Medicine

VOLUME 49

Editors

H. Tristram Engelhardt, Jr., *Center for Ethics, Medicine, and Public Issues, Baylor College of Medicine, Houston, Texas and Philosophy Department, Rice University, Houston, Texas*

Stuart F. Spicker, *Center for Ethics, Medicine, and Public Issues, Baylor College of Medicine, Houston, Texas*

Associate Editor

Kevin W. Wildes, S.J., *Department of Philosophy, Georgetown University, Washington, D.C.*

Editorial Board

George J. Agich, *School of Medicine, Southern Illinois University, Springfield, Illinois*

Edmund Erde, *University of Medicine and Dentistry of New Jersey, Camden, New Jersey*

Patricia A. King, J.D., *Georgetown University Law Center, Washington, D.C.*

E. Haavi Morreim, *Department of Human Values and Ethics, College of Medicine, University of Tennessee, Memphis, Tennessee*

The titles published in this series are listed at the end of this volume.

THE CODIFICATION OF MEDICAL MORALITY

Historical and Philosophical Studies of the Formalization of Western Medical Morality in the Eighteenth and Nineteenth Centuries

Volume Two: Anglo-American Medical Ethics and Medical Jurisprudence in the Nineteenth Century

Edited by

ROBERT BAKER
Department of Philosophy, Union College, Schenectady, New York

KLUWER ACADEMIC PUBLISHERS
Dordrecht/Boston/London

Library of Congress Cataloging-in-Publication Data

A C.I.P. Catalogue record for this book is available from the Library of Congress.

ISBN 0-7923-3528-7

Published by Kluwer Academic Publishers,
P.O. Box 17, 3300 AA Dordrecht, The Netherlands.

Kluwer Academic Publishers incorporates
the publishing programmes of
D. Reidel, Martinus Nijhoff, Dr. W. Junk and MTP Press.

Sold and distributed in the U.S.A. and Canada
by Kluwer Academic Publishers,
101 Philip Drive, Norwell, MA 02061, U.S.A.

In all other countries, sold and distributed
by Kluwer Academic Publishers Group,
P.O. Box 322, 3300 AH Dordrecht, The Netherlands.

printed on acid-free paper

All Rights Reserved
© 1995 Kluwer Academic Publishers
No part of the material protected by this copyright notice may be reproduced or
utilized in any form or by any means, electronic or mechanical,
including photocopying, recording or by any information storage and
retrieval system, without written permission from
the copyright owner.

Printed in the Netherlands

In Memory of

Freda Baker

TABLE OF CONTENTS

PREFACE

Like many novel ideas, the idea for this volume and its predecessor arose over lunch in the cafeteria of the old Wellcome Institute. On an afternoon in September 1988, Dorothy and Roy Porter, and I, sketched out a plan for a set of conferences in which scholars from a variety of disciplines would explore the emergence of modern medical ethics in the English-speaking world: from its pre-history in the quarrels that arose as gentlemanly codes of etiquette and honor broke down under the pressure of the eighteenth-century "sick trade," to the Enlightenment ethics of John Gregory and Thomas Percival, to the American appropriation process that culminated in the American Medical Association's 1847 *Code of Ethics*, and to the British turn to medical jurisprudence in the 1858 Medical Act.

Roy Porter formally presented our idea as a plan for two back-to-back conferences to the Wellcome Trust, and I presented it to the editors of the PHILOSOPHY AND MEDICINE series, H. Tristram Engelhardt, Jr. and Stuart F. Spicker. The reception from both parties was enthusiastic and so, with the financial backing of the former and a commitment to publication from the latter, Roy Porter, ably assisted by Frieda Hauser and Steven Emberton, organized two conferences. The first was held at the Wellcome Institute in December 1989; the second was sponsored by the Wellcome, but was actually held in the National Hospital, in December 1990.

Our plan was to publish the conferences as a two-volume set, in which each volume would explore a different century but would nonetheless be united to its companion by a single comprehensive conception reflected in the common title, "The Codification of Medical Morality: Historical and Philosophical Studies of the Formalization of Western Medical Morality in the Eighteenth and Nineteenth Centuries." We jointly edited volume one, *Medical Ethics and Etiquette in the Eighteenth Century*, which was published in 1993. By the time we turned our attention to the second volume, *Anglo-American Medical Ethics and Jurisprudence in the Nineteenth Century*, however, we found that the factors that had originally linked us – a common tie to the Wellcome Institute, the convenience of living in London, a common conception of our purpose – had dissolved. We were working at great distances from each other and were plagued moreover by delays. Graciously allowing the slowest to set the pace, Dorothy and Roy Porter turned over the task of completing the editing to me – although the overall conception of this volume was jointly theirs; they supplied most of the contributors with editorial comments, and, as noted earlier, Roy Porter was the actual organizer of the 1990 Wellcome conference. Perhaps the one point on which the present volume directly differs from the one we originally conceived is that it includes, as the original did not, three nineteenth-century codes of medical ethics (the *Boston Medical Police*, the unabridged 1847 AMA *Code of Ethics*, and Jukes Styrap's *Code of Medical Ethics*) and accompanying introductions.

In editing this volume I incurred many debts, particularly to the contributors for their willingness to revise the papers they presented at the conference – but most especially for their patience. I owe a special debt to two contributors in particular: Chester Burns, who although unable to attend the Wellcome Conference, nonetheless willingly provided a text of the 1808 Boston Medical Police and permitted us to reproduce his essay, "Reciprocity in the Development of Anglo-American Medical Ethics" (which is reprinted with the permission of Science History Publications); and Peter Bartrip, who supplied me with a copy of Jukes Styrup's code and who wrote a brief foreword.

Four great libraries provided the primary sources referred to throughout this volume: the American and Harry Ransom collections of the University of Texas at Austin; the Blocker History of Medicine Collection at the University of Texas Medical Branch at Galveston; the library of the New York Academy of Medicine; and, of course, the library at the Wellcome Institute of the History of Medicine, London. All the librarians were kind – although often bemused by the prospect of a philosopher searching for truth in archival sources – the most generous with her time, however, was undoubtedly Inci Bowman of the Blocker collection. Travel to these collections was funded by the Humanities Faculty Development Fund of Union College.

In closing, I should like to acknowledge two special debts: one to my friend and editor, Stuart Spicker, whose keen eye and incisive understanding brings out the best that a book has to offer; the other to my secretary, Marianne Snowden, without whose diligence, commitment and energy this book would not have come to fruition.

ROBERT BAKER

ROBERT BAKER

INTRODUCTION

In 1789, at the very birth of the United States, one of its founding fathers, Professor Benjamin Rush, signer of the Declaration of Independence and a founder of the medical college of the University of Pennsylvania, concluded his annual course with a lecture on the type of physician the new nation needed. He urged his students to be financially independent farmers who dealt honestly with their patients.

> There is more than one way of playing the quack. It is not necessary, for this purpose, that a man should advertise his skill, or his cures, or that he should mount a phaeton[1] and display his dexterity in operating to an ignorant and gaping multitude. A physician acts the same part, in a different way, who assumes the character of a madman or a brute in his manners, or who conceals his fallibility by an affected gravity and taciturnity in his intercourse with his patients. Both characters, like the quack, impose on the public ([36], p. 255).

Rush then urged his students to cultivate, in a new "American" way, the very virtues that had been inculcated in him by his teacher, John Gregory, of Edinburgh – piety, attention, humanity, and beneficence. He concluded with a stirring appeal.

> Human misery of every kind is evidently on decline. Happiness, like truth, is a unit. While the world, from the progress of intellectual, moral, and political truth, is becoming a more safe and agreeable place for man, the votaries of medicine should not be idle. All the doors of the temple of nature have been thrown open, by the convulsions of the late American Revolution. This is the time, therefore, to press upon her alters. We have already drawn from them discoveries in morals, philosophy, and government; all of which have human happiness for their object. Let us preserve truth and happiness, by drawing from the same source, in the present critical moment, a knowledge of antidotes to those diseases which are supposed to be incurable ([36], pp. 263–264).

R. Baker (ed.), The Codification of Medical Morality, 1–22.
© 1995 *Kluwer Academic Publishers. Printed in the Netherlands.*

Rush and his peers passed their moral fervor, their belief that old European medical ideals could be revitalized and reinvented in the newly formed United States, to their students, the next generation of American physicians. As the profession grew, and as municipal, county and state medical societies were formed, it became customary – commencing with the Boston Medical Police of 1808 [44] – for these societies to append to their constitutions "codes of medical ethics" drawn from the writings of John Gregory (1725–1773), Thomas Percival (1740–1804), and, of course, Benjamin Rush (1746–1813).

This codification effort culminated, fittingly, in 1847, with the formation of the American Medical Association, and the adoption of a national code of medical ethics. The physicians who founded the AMA believed that the code of ethics they had enacted would be an enduring legacy, at least as important as the organization they had founded. In the words of historian John Haller: "Following publication of the code in 1847, articles in medical journals across the country supported and elaborated upon [the code's] principles. . . . Doctors from Massachusetts to Texas drew from the code for innumerable speeches and lectures before their societies, graduating medical classes, and public lyceums. . . . [Some] enthusiastically claimed the code to be the most noble production of man since the Declaration of independence" ([21] pp. 237–238).

In recent years, however, "revisionist" sociologists Jeffrey Berlant [11], Paul Starr [38], and Ivan Waddington [41], [42], among others, have "revised" this lofty conception of the AMA's *Code of Ethics*. According to the revisionists – whose views are echoed in most standard histories of the nineteenth century American medicine [38] and medical ethics [15] – whatever the drafters themselves may have believed about their efforts, socio-historical analysis reveals these so-called "codes of medical ethics" to be nothing more than self-serving professional etiquettes. They have been dubbed "ethics" to disguise organized medicine's attempt to monopolize medical thought, so that, by driving homeopaths and other "irregular" competitors from the medical marketplace, it could ultimately monopolize medical practice. Crowning insult with injury, to quote Paul Starr, "while monopoly was doubtless the intent of the AMA's program, it was not the consequence. The irregulars thrived" ([38], p. 91).

Revisionist socio-historical critiques of codes of medical ethics actually wed a new methodology to old criticisms. From day one, physicians of all stripes, regulars and irregulars, had denounced the AMA's "monopolizing" intent, and had castigated the 1847 *Code of Ethics* as an "etiquette." One reason for this reaction, Robert Baker argues in his Introduction to the AMA's *Code* [5], is its language, its moral voice. The classic voice of medical morality, from the Hippocratic Oath to the lectures of John Gregory and Benjamin

Rush, is the first person singular. The AMA's *Code*, in contrast, is written in the second and third person plural; it is thus *not* an ethic of *personal* honor, written in the first person, as "*I*, " but an ethics of *professional* duties, rights and responsibilities. Not unnaturally, many physicians keenly resented the *Code's* submersion of their first person voices and the prerogatives of personal honor into a new, corporate, professional morality. In 1869, Alfred Carrol and John C. Peters, both physicians and editors of the *Medical Gazette* of New York, condemned the 1847 *Code* for ignoring the "personal sense of honor among the membership of an honorable profession" ([14] p. 238). Medical morality had traditionally been grounded in personal honor, so it seemed obvious to Carrol and Peters that any *Code* not so grounded was merely a presumptuous etiquette.

"Irregulars," on the other hand, raised a hue and cry over "monopolization." They had been stung doubly: not only had the *Code* excluded them from membership in the AMA, but physicians who consulted with them were under threat of expulsion. State and local societies affiliated with the AMA were also bound by these rules, threatening the "irregulars'" livelihoods because consultation was an essential feature of nineteenth-century medical practice. Forced to defend their legitimacy and their livelihoods, the "irregulars" fought back with every means at their disposal – including the pejorative "monopolization."

These epithets entered the mainstream of scholarly commentary in 1927 when Chauncey Leake, a professor of pharmacology, an avid reformer, a critic of the AMA – and a medical historian – reprinted an edition of Thomas Percival's *Medical Ethics* (1803) in conjunction with the AMA codes of 1847, 1903, and 1912. The conjunction was designed to demonstrate that American medical ethics rested on a semantic mistake; as Leake put it, "the term 'medical ethics', introduced by Percival, is really a misnomer" ([26] p. 1).

> [The AMA's 1847 *Code* is] based . . . on Thomas Percival's "Code," it refers chiefly to the rules of etiquette developed in the profession to regulate the professional contacts of its members *with each other*. . . . Unfortunately Percival was persuaded that "medical ethics" was the proper title for his system of professional regulations. All similar and subsequent systems of general advice, whether official or not have received the same title. As a result confusion has developed in the minds of many physicians between what may be really a matter of ethics and what may really be etiquette ([26] pp. 2–3).

Leake's analysis rests on two historical observations and one widely-accepted point of philosophical theory. The two historical observations are incontrovertible: first Percival had not originally intended to entitle his book "Medical Ethics" but was persuaded to do so by his friends ([31], p. 7); secondly, the

AMA's 1847 *Code* was based on Percival's *Medical Ethics*. The point of philosophical theory is also widely accepted and goes under the rubric "the applied ethics model" [9]. On this model medical ethics, in contrast to medical etiquette, is, to quote Leake, based upon "philosophical analyses of the principles of ethical theory made by recognized ethical scholars" ([26], p. 3). Since neither the AMA's *Code*, nor Percival's, is based on principles of ethical theory, they can not be ethics and so, Leake concluded, they merely reflected professional etiquette. Consequently, although "Percival sincerely and earnestly did his best to promote the idealism and dignity of the ancient profession of medicine . . . in failing to draw a clear distinction between points of etiquette between physicians . . . and matters of real ethical significance to humanity, Percival's 'Code' made it easy for professional task-masters to exact as severe a penalty for transgressions of one as for the other, so that now professional etiquette is often maintained at the expense of general morality" ([26], p. 57).

In the 1970s Berlant, Waddington, and other socio-historians rediscovered Leake's semantic analysis of the AMA codes, and incorporated it into a deeper analysis grounded in economic self-interest. Percival and other codifiers glorified professional etiquette as ethics, the revisionists argued, not because they were bamboozled by semantics, but because the real point of so-called "codes of medical ethics" was to establish a professional monopoly that protected regular practitioners against intellectual and economic competition. Leake's analysis was correct but naive: professional etiquette has been "maintained at the expense of general morality," not by accident, but by design – the principal function of professional codes of medical ethics is to disguise and justify the medical profession's financial stake in monopolizing medical practice.

Revisionist charges impeaching the integrity of eighteenth and nineteenth century codes of medical ethics are so serious that no history of the subject can ignore them. But how is one to assess their validity? Since their case depends heavily on Leake's analysis, suppose we reconsider his three points *seriatim*: that Percival changed the title of his book, that the AMA's 1847 *Code* was drawn from Percival, and that neither is justified in terms of fundamental ethical principles. Leake's claim that Percival was persuaded to change the title of his book to *Medical Ethics*, is correct. The original title had been, *Medical Jurisprudence*. The book's subtitle, "a Code of Ethics and Institutes Adopted to the Professions of Physic and Surgery," however, *never* changed and it clearly states that Percival believed himself to have written a code of *ethics*. Percival and his friends had been debating whether medical ethics was an independent subject, or whether it fell within the British medical jurisprudence tradition – which, as Chester Burns [12] and M. Anne Crowther [17] point out in their contributions to this volume, looks to the law for *moral* guidance. Percival's friends recognized that his work represented a significant departure from the medical jurisprudence tradition, and he ultimately acknowledged that they were right. The ethics-etiquette distinction was never discussed and is, moreover, significantly different from the ethics-jurisprudence distinction. Consequently, Leake is wrong, there was no semantic confusion, no misnomer, no conflation of ethics and etiquette,[2] to be read from the fact that Percival's friends prevailed upon him not to associate his book with the medical jurisprudence tradition. On the contrary, as John Pickstone has argued [32], a careful analysis, not only of Percival's titles and text, but also of the context in which Percival produced *Medical Ethics*, reveals that his undoubted intent was to write, and to be seen to be writing, a work that would perpetuate Enlightenment moral ideals in medicine.

Leake's second historical observation, while true, is also misleading. As Isaac Hays, the principal drafter of the AMA's 1847 *Code of Ethics* notes in his prefatory remarks, it was drawn from Percival's *Medical Ethics* ([22], p. 000). Nonetheless, in drafting the 1847 code, Hays and his colleagues freely altered Percival's precepts and framework when it suited them to do so. Percival believed that physicians' moral obligations arose out of a tacit compact between the profession and society [4]; however, the 1847 AMA *Code of Ethics* is structured as an explicit contract between physicians, society, *and* patients. Thus the American drafters transformed a *tacit* two-party compact (between society and profession) into an *explicit* tripartite contract (between society, profession, and *patients*). The drafters' reconceptualization of medical ethics in terms of reciprocal rights and responsibilities, moreover, is expressly noted by the chairperson of the drafting committee, Dr. John Bell, in his introduction to the *Code*: "Every duty or obligation implies, both in equity and for its successful discharge a corresponding right. As it is the duty of a physician . . . to expose his health and life for the benefit of the community, he has a just claim, in return, on all its members, collectively and individually" ([10] p. 66).

Unfortunately, the version of the AMA's 1847 *Code of Ethics* that Leake published is missing Bell's "Introduction" and thus lacks an explanation of the methodological underpinnings of the *Code*.[3] The most likely explanation for the abridgment is the difficulty of locating a complete version of the 1847 AMA *Code of Ethics*. The *Code*'s popularity and importance meant that dozens of abridged editions were published: The AMA itself published an abridged but undoubtedly official version of the code in every edition of its *Transactions* from 1857 to 1883; these, in turn, became the source of innumerable abridged editions that circulated under the misleadingly official title "*Code of Ethic of the American Medical Association, Adopted May 1847*" [2]. Moreover, the *Code of Ethics* predates both the AMA and its first official journal, the *Transactions*; consequently the unabridged *Code* could only be found in the few hundred copies of the *Minutes of the Proceedings of the National Medical Convention held in the City of Philadelphia, in May 1847* distributed to the delegates to the first official AMA meeting, held in 1848.

It was thus easy for Leake to accept an abridged version of the 1847 AMA *Code* as authoritative and complete, but it was also singularly unfortunate. Leake had differentiated between ethics and etiquette by arguing that the former rests on principles grounded in ethical theory, whereas the latter does not. Bell's "Introduction" explained the nature of the ethical theory that informed the 1847 *Code*; thus, from Leake's perspective, it explained the *Code*'s claim to ethical status. Reading the *Code* without it, he was left virtually clueless about the nature of its theoretical underpinnings and, quite naturally, construed it as a professional etiquette. Leake noticed, however, the one clue to the theoretical underpinnings of the *Code* that is clearly discernible in the abridged text: the titles of its three chapters. He remarks that "The chapter headings to the national 'Code' are interesting. . . . 'Of the Duties of the Profession to the Public and of the Public to the Profession.' Sweet conceit of the medical moralists of the 'Fabulous Forties!'"

Conceit, however, is in the mind of the beholder. Leake may have found an ethic of reciprocal relationships inconceivable, but the classic name for such an ethic is a "social contract." Leake's difficulty in recognizing the social contractarian claims of the 1847 *Code of Ethics* may arise because he accepted the applied ethics model: applied ethicists and contractarians look in different directions for moral justification. The standard representation of the applied ethics model is the following diagram from Tom L. Beauchamp and James Childress's textbook, *Principles of Biomedical Ethics* ([9], p. 6).

4. *Ethical Theories*
↑
3. *Principles*
↑
2. *Rules*
↑
1. *Judgments and Actions*

As Beauchamp and Childress explain: "According to this diagram, judgments about what ought to be done in particular situations are justified by moral rules, which in turn are grounded in principles and ultimately ethical theories" ([9], p. 6). The diagram thus graphically displays the "upward-looking" principled nature of the applied ethics model. As Beauchamp observes in his chapter on Worthington Hooker [8], on the applied ethics model, acts of confidentiality, truth-telling, and patient care are justified in terms of principles grounded in theoretical models.

Although some contractarian theorists (most notably, John Rawls [33]) use the model of the social contract to generate principles (in Rawls' case, famously, a principle of justice), the classic English contractarians (Thomas Hobbes and John Locke), and contemporary contractarians who take them as their models (David Gauthier [20] and Robert Nozick [27]), ground morality in the reciprocal nature of the social contract itself. Thus, to cite the title of one of David Gauthier's books, contractarian morality is "Morals By Agreement" [20]; there are no higher level principles or theories that justify contractarian morality. Justification stops with the contract itself.

Returning to Bell's "Introduction" to the *Code of Ethics*: whereas an applied ethicist would seek to justify the physician's "duty to expose his life for the benefit of the community" in terms of beneficence or some other moral principle, as a contractarian, Bell turned directly to reciprocity. Physicians, collectively undertook to treat patients "when pestilence prevails . . . even at the jeopardy of their lives," because ("Sweet conceit of the medical moralist of the . . . Forties!") they expected the public to be reciprocally bound to respect the profession. Bell and Hays thus had no need to justify their *Code of Ethics* by appealing to higher moral principles or to ethical theory; from their contractarian perspective reciprocal obligations sufficed in themselves. Leake's applied ethics conception of morality, however, demanded more; it demanded moral principles to differentiate ethics from etiquette. Consequently, when Leake read the abridged version of the *Code of Ethics* and found neither a principled justification, nor Bell's sketch of a theory of reciprocal obligation, he naturally dismissed the *Code* as etiquette.

The revisionist sociologists who predicated their theories on Leake's analysis did not interrogate his scholarship or probe his philosophical assumptions. They had no reason to do so: his conclusions conveniently supported their theories, and they were, after all, sociologists, not ethicists, historians, or philosophers. Their primary interests lay elsewhere: specifically, they were interested in refuting functionalism, the school of sociology – represented in America by Talcott Parsons [28], [29], and in Britain by A. M. Carr-Saunders and P.A.

Wilson [13] – that flourished in the 1950s and early 1960s. Functionalists believed that the characteristic features of fully-formed professions – altruism, autonomy, expertise, licensing, and professional codes of ethics – arise primarily because professionals sell a complex commodity whose quality a lay person is incapable of judging. Nonetheless, society desires to promote these commodities and, finding them too complex to regulate directly, it delegates its powers of social control to the professionals themselves – provided that they, in exchange, use this power to promote the public's good. Kenneth Arrow's economic analysis of the professions was contemporary with, and complemented, functionalist sociology. Arrow [3] argued that the professions represented a non-market societal response to the problems of imperfect competition; these non-market social institutions, in effect, fill the gap left by the inability of markets to perform their normal functions.

Functionalist sociological theories and classical economics, however, became unfashionable in the late 1960s, as scholars turned their attention, not to conditions that made society functional, but to those that made it dysfunctional. The slogan for rejecting the functionalist sociological theories and classical economics was that they rationalized the status quo by confusing the structural with the functional and the real with the rational. These rationalizations rendered functionalism incapable of comprehending the dysfunctional and irrational aspects of society. The problem with Parsons' analysis, Starr charged, was that: "[It] accepts the ideological claims of the profession...and ignores evidence to the contrary...[especially] the historical process that lies behind professional dominance" ([38] p. 21). Arrow's analysis is said to suffer from a similar defect: "The particular alternative to the competitive market that developed in America cannot be derived from a purely abstract analysis; it requires an analysis that is both structural and historical. . . . Arrow . . . attempts to explain the particulars of a system at a given moment in history in terms of the universal features of medicine" ([38], p. 227). Berlant too rejected functionalism, "because of its virtually non-existent historical grounding" ([11], p. 300). On the other side of the Atlantic, Waddington was offering precisely the same argument: functionalist analyses failed when subjected to examination in the actual historical contexts [41], [42]. Thus history, especially the history of medical ethics, was to provide the laboratory in which the revisionist sociologists would demonstrate the flaws of functionalism.

"Part of the historical task," Starr wrote, "is to explain . . . how various . . . claims to legitimacy became established, how they took institutional form, how the boundaries of medical authority expanded, and how authority translated into economic power and influence" ([38], p. 16). In writing this history it was important that traditional "professional claims not be taken simply at face value . . . [but] should be seen as means of legitimating professional authority, achieving solidarity among practitioners and gaining a grant of monopoly from the state." ([38], p.15). I dub methodological principle that Starr enunciates here – the principle that professional claims should always be read in terms of professional ambitions – the "discounting rule" because it is used to preemptively discount professionals own accounts of their reasons and motives. The discounting rule is an exceptionally powerful methodological assumption precisely because it permits the revisionists to preemptively dismiss anything physicians themselves say about their intentions, reasons, or motivations – *a priori* – except when they support revisionist theories. The net effect of the discounting rule was to enable sociologists to project their theories onto the history of medicine without being "encumbered" by the accounts offered by the historical actors themselves.

Once the revisionists had discounted the actual statements of all those who drafted or endorsed codes of ethics, they built their analysis on the hypotheses that professional medical ethics are (1) etiquettes, designed to disguise a professional drive to (2) monopolize the medical marketplace. The first part of their analysis rests heavily on Leake's scholarship, which, as we observed, does not withstand serious scrutiny. Moreover, as Stanley Reiser's comprehensive content analysis of the 1847 AMA *Code of Ethics* demonstrates, the 1847 AMA *Code* embraces the major themes of medical ethics, past and present [34]. Leake's etiquette theory appears, in retrospect, to have been an artifact of unfortunate happenstance exacerbated by misleading preconceptions; the theory is best placed in the dust bin of discarded ideas.

The second line of argument supporting the reductionist case – the monopolization hypothesis – can not so readily be discarded. The theory rests on overwhelming evidence that nineteenth-century medical societies in America and Britain (i) attempted to secure a state licensing system that (ii) limited practitioners to those who had an approved education in medical science; that (iii) the societies also prohibited advertising and (iv) attempted to regulate fees; that (v) they tried to expel and delegitimate practitioners who did not abide by their strictures on education, licensing, and advertising; and (vi) that they tried to expel members who consulted with practitioners who were ineligible for membership. These practices appear to support the monopolization hypothesis since (i) and (ii) restrict the supply of physicians, thereby raising the price of their services; (iii) is anti-competitive on its face, and destroys the prospect of a market; (iv) is essentially price-fixing; while (v) and (vi) constitute classic monopolistic attempts to eliminate the competition by non-market mechanisms.

The monopolization theory comes in two forms: strong and weak. Starr and Waddington embrace the stronger form of the theory: that monopolization is the driving mechanism underlying professionalization and that the *sole* function of codes of medical ethics is to serve as a fig leaf disguising the monopolizing designs of the profession. Without the support of Leake's etiquette theory, the strong "fig leaf" theory takes on a radical edge; for, if there is no difference between the 1847 *Code of Ethics* and other forms of professional ethics, holders of the strong monopolization theory are committed to arguing that all professional ethics – perhaps even all forms of ethics – are mere fig leaves for individual and collective interests.

The weaker variant of the theory, the form endorsed by Berlant, has less radical implications. Berlant allows that professional ethics *may* have the moral function of "help[ing] protect or further the medical interests of patients" ([11], p. 125), and that this *may* have been "the intent of the creators of medical ethics"; nonetheless, he argues, "the creators of medical ethics were aware of the possibility of monopolization when they wrote monopolistic ethics," ([11], p. 126) and these codes were later used to further monopolistic ends. Thus Berlant, unlike Starr and Waddington, opens the door to the possibility that the nineteenth-century codes of medical ethics *may* have purposes other than monopolization. His point is that, irrespective of whether they were intended to serve, or actually served, any moral function, they were undoubtedly intended to serve, and actually served as mechanisms of monopolization.

The fundamental weakness of all monopolization theories is that they presume the possibility of a free market. As Kenneth Arrow [3] and virtually all medical economists agree (Milton Friedman is the exception [19]), there is not a free market for medicine because market transactions are limited by the purchaser's ignorance and vulnerability (see the case of Abigail Plumer, in Chapter One), and by the seller's reluctance to refuse treatment. Since neither buyer nor seller is acting freely, the market for medical care can be neither free nor competitive – and is thus not really a market. It is, therefore, nonsense to treat the six constraining practices cited by revisionists as clear evidence of monopolization; they are just as easily interpreted as efforts to remedy the imperfections of the market.

The imperfections of the medical marketplace were well understood, by nineteenth-century patients: consider the following remarks penned by Joseph G. Baldwin in 1853.

Nobody knew who or what they were, except as they claimed, or as a surface view of their characters indicated. Instead of taking to the highway and magnanimously calling upon the wayfarer to stand and deliver . . . some unscrupulous horse doctor would set up his sign as "Physician and Surgeon" and draw his lancet on you, or fire at random a box of pills into your bowels, with a vague chance of hitting some disease, unknown to him, but with a better prospect of killing the patient, whom or whose administrator, he charged some ten dollars a trial for his marksmanship ([25], pp. vii–viii).

Baldwin here calls our attention to a major imperfection of the unregulated medical marketplace: "Nobody knew who or what they were, except as they claimed." His implicit demand for some form of licensing or registration confirms Arrow's analysis that such demands are attempts to correct market imperfections, in this case by informing purchasers about providers. Thus the demand for licensing – the evidence most commonly cited by revisionists to support the monopolization hypothesis – is better read as evidence for Arrow's theory that professionalization is a mechanism for correcting the imperfections of the medical market.

The AMA *Code*'s proscription of secret nostrums and the prohibition of advertising can also be interpreted as attempts to correct market imperfections. In normal markets advertising conveys information that consumers can use to make rational choices. In the medical market, however, even if purveyors inform purchasers truthfully about the contents of their wares, consumers characteristically lack the ability to understand their physiological and pharmacological implications. Advertising is thus more readily used to defraud than to inform, and the only way to protect the consumer is either to regulate it (as we now do in America and Britain), or to prohibit it entirely (the solution proposed by the AMA's *Code*).

The same line of argument establishes fee-scales as a prerequisite for the medical market: for, to reiterate Arrow's thesis, in the medical market consumers are characteristically unable to assess the value of the services rendered. They are thus better off if price-competition is prohibited, and if competition is restricted to areas consumers can assess for themselves – such as the responsiveness of providers.

The point to appreciate here is not that Arrow's case has been proven. It is rather that massive evidence cited by the revisionists about licensing, fee-setting, prohibitions on advertising, and so forth *does not, and can not prove their case*, since it is as compatible with Arrow's analysis (and thus with functionalist and contractarian analyses) as it is with their own. More fundamentally, however, *unless* the revisionists can disprove Arrow's contention that there is no free market in medicine, their monopolization hypothesis is meaningless. Monopolization presupposes the possibility of a free market, if there is no possibility of such a market in medicine, the expression, "monopolization of the medical marketplace," quite literally, has no meaning. In the absence of a market, talk of "monopolization" is nonsense.

The monopolization hypothesis also fails a fundamental test of historical explanation. Good historical explanation renders the past comprehensible to the present. Revisionists abide by this precept when they attempt to explain the actions of the nineteenth-century physicians in terms of economic self-interest – a motive well understood by people in the present. Yet, as the revisionists themselves remark, codes of medical ethics never served the economic interests of nineteenth-century physicians. According to Starr, "the ethical code itself exacerbated divisions because it excluded sectarian physicians" ([38], p. 94). Monopolization requires political power: had the AMA's aim been monopolization, its best strategy would have been to incorporate *all* present practitioners, from every school of practice, into their organization. Once this was achieved, they could then begin to control the market by preventing price competition, setting high standard fees, and limiting each practitioner's case load (creating artificial scarcity and thus raising the price of their services). Instead, they alienated their potential allies – homeopaths and other sectarians – by excluding them from the AMA and the affiliated state societies. They then compounded their error by alienating the medical schools by insisting on a scientific pre-medical education, a lecture-based scientific medical education and supplementary clinical experience ([35], pp. 108–121). If the AMA's objective really was monopolization, its actions were remarkably ill-conceived.

A different and more reasonable picture of the actions of nineteenth-century medical societies emerges if one rejects the revisionists' discount rule and considers seriously what the founders of the American medical societies wrote and said about their motives and reasons. Historian Joseph Kett, writing in an era before the discount rule became the *modus operandi* for social historians of medical ethics, provides the following account of the behavior of the physicians in New York, South Carolina, and Ohio medical societies, during the years following the passage of the 1847 AMA *Code*.

> Their aims were lofty. They sought improvements of preliminary education, and tighter regulation of medical ethics, especially in relationship to consultations with quacks. They besought their members to write scholarly papers and to keep up with medical literature. They denounced members of the medical profession who advertised nostrums. In a host of ways they sought to make the profession more useful and hence more esteemed. By 1860 the voluntary associations had generally committed themselves to the ideal of a self-regulated profession ([25], p. 176).

Notice that, with one singular exception, Kett's careful and detailed analysis of state medical societies coincides with that offered by revisionists. The exception, of course, is the discount rule: Kett takes the society members at their word and thus characterizes their aims as "lofty." The revisionist, by contrast, castigate these declarations as hypocritical statements uttered to disguise a drive for monopolization.

Which account is more plausible? Kett and the revisionists agree that for the most part the societies were unable to achieve their aims during the nineteenth century. The difference is that Kett can offer an explanation of the members futile efforts, while the revisionists can not.

> The gap between expectation and achievement was not traceable to apathy or dilatoriness on the part of leaders. There was little joy in traveling a few hundred miles to an annual convention and then listening to the profession excoriated by its members. A strong sense of obligation was rooted in the conscience of antebellum physicians.
>
> The problem was not moral delinquency but a fundamental flaw in the idea of a voluntary association. On a theoretical level the voluntary societies were trying to move in two directions at once. . . . The voluntary societies were . . . trapped in a dilemma. Either they kept their membership requirements loose, in which case they could hardly have claimed to have purified their ranks, or they tightened requirements and lost any chance of presenting a united front ([25], p. 177).

The picture that Kett paints here is painfully familiar: nineteenth-century medical society members were caught, as many of us often are, between their ideals, "their strong sense of obligation," and their economic self-interest; between their codes of ethics, and their economic interest in seizing political power. This conflict was exacerbated when states weakened licensing requirements in the mid-eighteen-forties and it consumed much of the American medical profession during the second half of the century. Yet, from the revisionist perspective, this conflict – one of the predominant features of nineteenth-century American medical life – was inexplicable. It should not have existed: codes of medical ethics should not have conflicted with physicians' economic interests since, according to the revisionists, the codes were empty rhetoric that served only to further economic interests

What were the lofty aims that prevented medical associations from opening their doors to all comers? As Kett notes, it was their commitment to scientific medicine, to a scientific medical and pre-medical education, that forced them into a reformist and exclusionary stance. The reformist stance is evident, not only from John Bell's introduction to the AMA's 1847 *Code of Ethics* – which had called on physicians to be "trustees of science and almoners of benevolence" and thus to "prevent the introduction into their body of those who have not been prepared by a suitable preparatory moral and intellectual training" – it is also the theme struck by the first President of the AMA, Professor Nathaniel Chapman of the University of Pennsylvania (mentor to both Bell and Hays), when he welcomed the delegates to the first annual meeting of the American Medical Association held in Baltimore, May, 1848.

> This assemblage presents a spectacle of moral grandeur delightful to contemplate. Few of the kind have I ever witnessed more imposing in its aspect, and certainly none inspired by purer motives, or having views of a wider range of beneficence. The profession to which we belong . . . has become corrupt, and degenerate, to the forfeiture of its social position, and with it the homage it formerly received spontaneously and universally. . . . [Is not] the profession . . . environed by difficulties and dangers, arising mainly from the too ready admixture into it of individuals unworthy of the association, either by intellectual culture, or moral discipline, by whom it is abased? And are you not imperatively instructed to purify its taints and abuses and restore it to its former elevation and dignity? ([16], pp. 8–9).

As Kett observes [25], and as is manifestly clear from the texts of both Bell's Introduction and Chapman's welcoming remarks, the founders of the American Medical Association and the drafters of its *Code of Ethics* were not self-abnegating moralists. They believed in struggle, but not for its own sake; if they embraced the virtues of morality and science, they fully believed that in the end society would reward the medical profession for doing so. Beneficence and science, while good for their own sake and for the sake of patients, were also good for the profession and its members. This was the beauty of the tripartite contractual relationship they envisioned: physicians virtues would be rewarded by societal prestige and patients' respect. Yet to earn this prestige and respect, the profession needed to commit itself to science. They had to bring the theories and practices of European medical science to the American continent; they had to introduce Auenbrugger's percussion, Bichat's autopsies, Corvisart's methods of clinical observation, Laennec's stethoscopy and Louis's statistics into the mainstream of American medical practice. In its early years the AMA formed implementation committees to do precisely this – parceling out the task in terms of special committees for anatomy, chemistry, forensic medicine, hygiene and sanitary measures, medical science, physiology, materia medica, surgery, obstetrics, publications, vital statistics – and, most importantly, medical education.

If American physicians were to be scientific, they must be educated in medical science. Indeed, the founders of the AMA were so profoundly committed to this ideal that they challenged both the apprenticeship tradition of training physicians and surgeons, and the medical schools. They challenged not only fly-by-night proprietary schools and diploma mills (like the notorious Willoughby Medical College of Ohio, which issued a fifth of its diplomas as honorary degrees – that is degrees awarded to those who had neither attended lectures nor passed examinations, ([25] p. 175)) but also the most prestigious medical schools in the nation. These, they charged, were better at collecting fees than at providing laboratory space, or clinical experience, or examining students, or even presenting an integrated course of lectures. In 1848, the AMA asked medical colleges to provide their students with laboratory demonstrations, clinical instruction, and a full *six* months of lectures on which they were *examined*. The leading members of the Harvard medical faculty – John Ware, Jacob Bigelow, and Oliver Wendell Holmes – refused:

> It is not expedient to extend the course of medical lectures beyond four months, as recommended by the American Medical Association. . . . The whole propositions proceeds from what seems to us a strange exaggeration of the importance of teaching by lectures, as compared with the other means of medical instruction. . . . No course of lectures, however prolonged, can give complete instruction in any department of professional study. If this be attempted, the teaching must necessarily be superficial. Lectures cannot communicate all the knowledge of the profession; they cannot approach this. It is a mistake to think of lectures as having this purpose. Their real value is diminished, where they are given with such a view, of their proper object. The great purpose of lectures should be, to teach the student how to learn for himself. . . . Learning is a thing which no man can do for another; the weight of education must fall on the learner; what he does not get and make his own, by the active exercise of his own powers he does not get at all. . . .
>
> We feel compelled, therefore, to express our decided conviction that four months in the year is quite as large a proportion of the student's time as can be profitably occupied in attendance on lecture. . . .
>
> We also [reject] clinical demonstrations and examinations because they would result in doubling a burden which is now as great as students can bear without [their] failing in health ([43], pp. 353–357).

Why were the founders of the AMA so insistent on lectures and demonstrations when mid-nineteenth century medical science offered practitioners so very little? They were not ideologues: Austin Flint struggled throughout his life to distinguish between typhus and typhoid; Isaac Hays (who drafted the *Code of Ethics*) had critiqued Bichat; Alfred Stillé (first Secretary to the AMA) struggled for years to develop a theory of contagion – but that achievement had to wait upon the work of Pasteur and Koch. The founders knew all too well the limits of the science of their age. What drove them to challenge the most prestigious medical colleges in the land was that they also knew that medicine taught entirely by apprenticeship, medicine learned without lectures or scientific demonstrations and without clinical experience, could not be scientific. What drove them was not the certainty that they knew the scientific truth, but the certainty that American medicine could never become scientific, could never attain Benjamin Rush's goal of attaining "a knowledge of antidotes to those diseases which are supposed to be incurable," unless American medical and pre-medical education was firmly grounded in science. They were not, as the revisionists have observed, initially successful. But with the opening of Johns Hopkins Medical School in 1893 and the radical reform of medical education catalyzed by the Flexner report of 1910, their ideas were ultimately triumphant.

The same line of analysis led the founders of the AMA and affiliated medical societies to reject sectarian medicine. They knew all too well that they lacked remedies for most ailments. Oliver Wendell Holmes freely admitted that "if the whole *materia medica, as now used*, could be sunk to the sea, it would be all the better for mankind – and all the worse for the fishes" ([23], p. 203). Yet in "Homeopathy and its Kindred Delusions" ([23], pp. 101–102), Holmes delivered a scathing critique of homeopathy, condemning it for seeking truth, not through scientific inquiry, but through the revelations of its founder, Samuel Christian Freidrich Hahnemann (1755–1843). Holmes argued that everyday medical experience disconfirmed the fundamental principle of homeopathic theory, the law of similars (*similia similibus curantur* – what causes illness in a healthy person will cure the same illness in a sick person); and he charged that homeopathic therapeutics – which called for the dilution of potent agents to less than 1/1,000,000 of a drop – violated the laws of chemistry. Worthington Hooker, a staunch defender of the AMA's 1847 *Code*, charged Hahnemann with succumbing to the *post hoc ergo propter hoc* fallacy (literally, "after this, therefore, because of this"; the fallacious confusion of succession with causality), "the *modus operandi* by which the genius of imposture is produced from the fantastic and ever-changing shapes of empiricism" ([24], p. vii). Nathan Smith Davis – the physician whose concerns about medical education initiated the sequence of events that culminated in the formation of the AMA – argued that "exclusive individual schools of medicine, founded on some one universal law of disease or equally universal law of cure" were incompatible with "strictly scientific investigations, clinical experience, and free discussions" ([18], p. 175).

The founders of the American Medical Association thus appear to have rejected sectarian medicine for principled reasons. They did this knowing that, as many commentators have since observed, it was against their economic and political self-interest to do so. Why then did they do it? The most reasonable explanation is that they believed their own rhetoric – they really were reformers.

Ironically, although the AMA's vision of a profession elevated by a dual commitment to ethics and science was prescient, its policy of exclusion, unlike its policies on scientific education, were not directly vindicated by events. Homeopathy did not whither away; it flourished, but in a way that was inconceivable to Bell, Chapman, Davis, Hays, Hooker and other founding members of the AMA. Bowing to the dynamism of nineteenth-century medical science and recognizing the legitimization it was conferring on conventional medicine, homeopaths eclectically incorporated conventional medical science into their schools. As Haller observes:

> By the 1880s and 1890s the requirements for graduation from the Hahnemann Medical College of Philadelphia were remarkably similar to most orthodox medical schools of the day. The three year curriculum included courses on anatomy, physiology, chemistry, surgery, therapeutics, pharmacy and toxicology. The books used for its courses were identical with those used in the schools of the regular [and included some written by Stillé and other founders of the AMA]. Not surprisingly, many homeopaths, upon graduation, declined to connect themselves publicly with the doctrines of Hahnemann and refused even to use the name homeopath ([21], p. 125).

Homeopathy survived, and eventually assimilated into the mainstream of American medicine, but it did so only by affirming the scientific basis of medicine – that is, it survived by surrendering to the conditions laid down by the founders of the AMA.

If the motives that the AMA's founders attributed to themselves provide a more coherent picture of the period than the revisionist monopolization hypothesis, what remains of the revisionist reading of nineteenth-century medical ethics? One could argue that some small something remains, that the founders of the AMA and the members of its affiliated societies were perhaps a bit more self-interested, a bit more concerned with economics, than they admitted. Oddly enough, this ultra-weak version of the monopolization hypothesis serves little purpose when discussing nineteenth-century medical societies. They were remarkably frank about recognizing the congruence of public, professional, and personal interests. If they endorsed science as the basis of medicine, if they stood for educational reform, if they accepted moral responsibility, they understood that in so doing they were serving not only the public and the profession, but also themselves as professionals. In their minds, public, private and professional goods were one: they understood and forthrightly professed that by acting in the public's interest, they were also acting in their own.

****

The papers collected in this volume, like those in its predecessor, are united by a common skepticism about the tenets of revisionism; they treat nineteenth-century codes of medical ethics seriously, as ethics. Since many readers may be unfamiliar with these codes, each of the two parts of the book opens by reproducing the most important American and British nineteenth-century codes of ethics – prefaced by an introductory essay setting the code in its historical context. Part One, which focuses on America, opens with the first American code of medical ethics, the Boston Medical Police of 1808 (introduced by Robert Baker). The next chapter contains an unabridged version of the 1847 AMA *Code of Ethics* (also introduced by Baker). Chapter Three is a detailed content analysis of the 1847 *Code of Ethics* by Stanley Reiser [34]. In Chapter Four, Tom L. Beauchamp analyzes Worthington Hooker's critique of the AMA *Code*'s paternalism.

In Chapter Five, Robert Veatch reminds us that there is more to nineteenth-century medical ethics than codes drafted by physicians; moral theology also offered an ethical perspective on the practice of medicine. Moreover, as Veatch demonstrates by comparing the AMA's *Code of Ethics* with nineteenth-century Catholic medical ethics, the two approaches to medical ethics were often strikingly different. Nor were Americans the only physicians concerned about medical ethics; in Chapter Six, Chester Burns' careful chronological analysis of the Anglo-American dialogue on medical ethics, shows that while it was British physicians – John Gregory (1724–1873), Thomas Percival (1740–18O4), and Michael Ryan (1800–1841) – who initiated the discussion of medical ethics, in the nineteenth century the impetus shifted to America, starting with the Boston Medical Police of 1808 and culminating in the AMA code of 1847.

This background sets the stage for Part Two, nineteenth-century Britain and the discussion of a code that seemingly failed to shape nineteenth-century British medical ethics – the code that the British Medical Association never adopted. After the BMA thrice failed to approve a code of ethics, the primary drafter of a British code, Jukes Styrap, published his *A Code of Medical Ethics* on his own. Yet, as Peter Bartrip remarks in his Introduction, even though Styrap's *Code* was never officially adopted by the BMA, it nonetheless set unofficial standards for the "done thing" and became the accepted, albeit unofficial, guide to proper medical behavior.

Why did the medical societies of Britain eschew official codes of medical ethics? The revisionist reply is that they had no need for such a code because they were granted a monopoly on licensing by the 1858 Medical Act, which set up an official *Medical Register* of all licensed physicians. This explanation provides a cogent answer to the question of why the BMA rejected Styrap's offer of a code in 1882. It does not explain why official codes of ethics were consistently rejected in the 1830s and 1840s. A more compelling answer is offered in Chapter Eight by M. Anne Crowther, who points to the deeply rooted British tradition of grounding medical ethics in jurisprudence – a tradition epitomized, she argues, in *Taylor's Principles and Practice of Medical Jurisprudence* [39], first published in 1844, and still published today. Chester Burns remarks the same tradition in Michael Ryan's 1831 [37] manual of medical jurisprudence, and Percival's 1794 publication, *Medical Jurisprudence*. The jurisprudence tradition looked to the law, not only to delimit unacceptable practice, but, by so doing, to implicitly define the acceptable and the ideal practice. Thus, while it was entirely consonant with the American spirit of private endeavor for a private association, the AMA, to appropriate the text of the first two chapters of Percival's 1803 *Medical Ethics* as the basis of its code of ethics, it was equally consonant with the British tradition, and with the entire text of Percival's original 1794 work, *Medical Jurisprudence*, to eschew private codes of ethics and to search the law for societal expectations about proper and improper medical conduct.

By embracing the medical jurisprudence tradition, however, the British made the task of the historians of medical ethics even more difficult, for there were no formal codes that laid out official moral standards. Thus, as Peter Bartrip notes in his careful analysis of the editorial policy of the *British Medical Journal*, an official publication of the BMA, historians need to look to medical journals and other sources to discern both the avowed standards of the profession, and the seriousness with which professionals treated these avowals. Bartrip points out that from the 1870s onwards the BMJ was caught in a moral dilemma: its editorial pages denounced the vice of secret remedies, even as its advertising columns were pronouncing their virtues. A straightforwardly revisionist reading would predict that since ethical avowals are not seriously intended, the BMJ would simply ignore this inconsistency – or jettison its principle. In reality, however, the editors did neither: once they realized that they were caught between their avowed principles and their purses, they struggled to untangle themselves and to develop a workable compromise.

In the last chapter of this volume, Russell Smith deals with another artifact of the British medical jurisprudence tradition, the General Medical Council (GMC). The Medical Act of 1858 set up an official *Register* of all official medical practitioners. The GMC's function was to police the *Register* by setting the educational standards that practitioners had to satisfy in order to be listed, and by determining which practitioners engaged in conduct so clearly unprofessional that they were to be stuck off the *Register*. The latter duty requires a determination of the minimal acceptable standards of conduct; it thus establishes a "minimalist" code of conduct. Smith traces the evolution of this minimalist code from the first edict issued by the GMC to relatively recent times, chronicling in some detail the problems facing a disciplinary body that attempts to uphold minimal moral standards of conduct without recourse to a formal code of ethics.

The contributors to this volume seem to tell a cohesive story about how the Enlightenment medical ethics that Gregory and Percival bequeathed to the English-speaking world came to be codified as medical ethics in America and as medical jurisprudence in Britain. They did not, of course, set out to tell a cohesive story; if they appear to have done so, it is probably because, for the first time in decades, the history of nineteenth-century Anglo-American medical ethics has been freed from the methodological fetters of revisionism.

Union College
Schenectady, New York
U.S.A.

NOTES

[1]A light four-wheeled horse-drawn carriage that has one or more seats facing forward.

[2]Moreover, Baker [4] has cited instances in which Percival distinguishes between ethics with etiquette in the text of *Medical Ethics*, thereby demonstrating that he did *not* conflate these two concepts.

[3]As Tom Beauchamp remarks in end note 5, p. 117, Leake's edition of Percival's *Medical Ethics* "is not entirely reliable," because it omitted and abridged many small but nonetheless significant parts of Percival's text – without informing readers. It is possible, given Leake's scholarly style, that he was aware of Bell's Introduction to the *Code of Ethics*, but, deeming it irrelevant, decided not to publish it.

BIBLIOGRAPHY

1. American Medical Association: 1847, "Code of Ethics," *Minutes of the Proceedings of the National Medical Convention held in the City of Philadelphia, in May 1847*, pp. 83–106; this volume, pp. 75–88.
2. American Medical Association: 1867, *Code of Ethics of the American* Medic*al Association, Adopted May 1847*, J. Windterburn, San Francisco.
3. Arrow, K.: 1963, "Uncertainty and the Welfare Economics of Medical Care," *American Economic Review* 53, 941–69.
4. Baker, R.: 1993, "Deciphering Percival's Code" in [6], pp. 179–212.
5. Baker, R.: 1995, "The Historical Context of the American Medical Association's 1847 *Code of Ethics*, this volume, pp. 47–64.

6. Baker, R., Porter, D., and Porter, R.: 1993, *The Codification of Medical Morality: Historical and Philosophical Studies of the Formalization of Western Medical Morality in the Eighteenth and Nineteenth Centuries: Volume One: Medical Ethics and Etiquette in the Eighteenth Century*, Kluwer Academic Publishers, Dordrecht.
7. Baldwin, J.: 1853, "Flush Times of Alabama and Mississippi" in [25], pp. vii–viii.
8. Beauchamp, T. : 1995, "Worthington Hooker on Ethics in Clinical Medicine," this volume, pp. 105–120.
9. Beauchamp T. and Childress, J.: 1989, *Principles of Biomedical Ethics*, 4th ed., Oxford University Press, New York.
10. Bell, J. 1847, "Introduction to the Code of Medical Ethics" *Minutes of the Proceedings of the National Medical Convention held in the City of Philadelphia, in May 1847*, pp. 83–92; this volume, pp. 62–72.
11. Berlant, J.: 1975, *Profession and Monopoly: A Study of Medicine in the United States and Great Britain*, University of California Press, Berkeley.
12. Burns, C.: 1977, "Reciprocity in the Development of Anglo-American Medical Ethics, 1765–1865," in Burns, C. (ed.), *Legacies in Ethics and Medicine*, pp. 300–6, Science History Publications, New York; this volume, pp. 133–144.
13. Carr-Saunders, A. and Wilson, P.: 1933, *The Professions*, Clarendon Press, Oxford.
14. Carrol, A. and Peters, J.: 1869, "Code or Ethics," *Medical Gazette* III, 150.
15. Chapman, C.: 1984, *Physicians, Law and Ethics*, New York University Press, New York.
16. Chapman, N.: 1848, "Presidential Address," *Transactions of the American Medical Association* 1, 8–9.
17. Crowther, M. A.: "Forensic Medicine and Medical Ethics in Nineteenth Century Britain," this volume, pp. 173–190.
18. Davis, N.: 1903, *History of Medicine, with the Code of Medical Ethics*, Cleveland Press, Chicago.
19. Friedman, M.: 1962, *Capitalism and Freedom*, University of Chicago Press, Chicago.
20. Gauthier, D.: 1986, *Morals by Agreement*, Oxford University Press, Oxford.
21. Haller, J.: 1981, *American Medicine in Transition 1840–1910*, University of Illinois Press, Urbana, IL.
22. Hays, I.: 1847, "Note" to *American Medical Association Code of Ethics*, this volume, pp. 73–74.
23. Holmes, O.: 1891, *Medical Essays, 1842–1882*, Houghton Mifflin, Boston.
24. Hooker, W.: 1849, 1972, *Physician and Patient*, Arno Press (reprint), New York.
25. Kett, J.: 1968, *The Formation of the American Medical Profession: The Role of Institutions, 1760–1860*, Yale University Press, New Haven.
26. Leake, C.: 1927, *Percival's Medical Ethics*, Williams and Wilkins, Baltimore.
27. Nozick, R.: 1974, *Anarchy, State and Utopia*, Basic Books, New York.
28. Parsons, T.: 1951, *The Social System*, Free Press, Glencoe, IL.
29. Parsons, T.: 1958, "Definitions of Health and Illness in the Light of American Values and Social Structure," in Jaco, E. (ed.), *Patients, Physicians and Illness*, pp. 165–87, Free Press, Glenco, IL.
30. Percival, T.: 1794, *Medical Jurisprudence or a Code of Ethics and Institutes Adopted to the Professions of Physic and Surgery*, privately circulated, Manchester.
31. Percival, T.: 1803, *Medical Ethics; Or, A Code of Institutes and Precepts, Adapted to the Professional Conduct of Physicians and Surgeons*, J. Johnson, London.

32. Pickstone, J.: 1993, "Thomas Percival and the Production of *Medical Ethics*," in [6], pp. 161–78.
33. Rawls, J.: 1971, *A Theory of Justice*, Harvard University Press, Cambridge, MA.
34. Reiser, S.: 1995, "Creating a Medical Profession in the United States: The First Code of Ethics of the American Medical Association," this volume, pp. 89–104.
35. Rothstein, W.: 1972, *American Physicians in the Nineteenth Century: From Sects to Science*, The Johns Hopkins University Press, Baltimore.
36. Rush, B.: 1789, "Observations on the Duties of a Physician, and the Methods of Improving Medicine. Accommodated to the Present State of Society and Manners in the United States," in Rush, B.: 1815, *Medical Inquiries and Observations*, Vol. 1, pp. 251–64, M. Carey, Philadelphia.
37. Ryan, M.: 1831, *A Manual of Jurisprudence, compiled from the best medical and legal works: comprising an account of: The Ethics of the Medical Profession, II. The Charter and Statutes Relating to the Faculty; and III. All Medico-legal Questions, with the latest discussions. Being an Analysis of a Course of Lectures on Forensic Medicine Annually Delivered in London and intended as a compendium for the use of barristers, solicitors, magistrates, coroners, and medical practitioners*, Renshaw and Rush, London.
38. Starr, P.: 1982, *The Social Transformation of American Medicine*, Basic Books, New York.
39. Taylor, A.: 1844, *A Manual of Medical Jurisprudence*, John Churchill, London.
40. Veatch, R.: 1995, "Diverging Traditions: Professional and Religious Medical Ethics of the Nineteenth Century," this volume, pp. 121–134.
41. Waddington, I.: 1975, "The Development of Medical Ethics – A Sociological Analysis", *Medical History* 19, 36–51.
42. Waddington I.: 1984, *The Medical Profession in the Industrial Revolution*, Gill & Macmillan, Dublin.
43. Ware, J., Bigelow, J., and Holmes, O.: 1849, "Letter to American Medical Association," *Transactions of the American Medical Association* II, 353–7.
44. Warren, J., Hayward, L., and Fleet, J.: 1808, *The Boston Medical Police*, Association of Boston Physicians, Boston; this volume, pp. 41–46.

PART ONE

THE NINETEENTH-CENTURY AMERICAN CODIFICATION OF MEDICAL ETHICS

CHAPTER 1

ROBERT BAKER

AN INTRODUCTION TO THE BOSTON MEDICAL POLICE OF 1808

Revisionists charge that professional codes of medical ethics amount to little more than trade union regulations wrapped in a fig-leaf of etiquette and puffed up with elevated rhetoric. No code lends itself more readily to this interpretation than the Boston Medical Police of 1808. Carleton Chapman, an eminent physician and historian of medicine, discounts the ethical significance of the Boston Medical Police in a single dismissive paragraph.

> *Percival's Ethics Crosses the Atlantic.* The American Revolution had not long settled when a standing committee of the Association of Boston Physicians cited Percival's magnum opus, along with works by Benjamin Rush and John Gregory, the committee having been instructed in 1807 "to propose a Code of Medical Police." The result was a short document . . . which was accepted by the Association early in 1808. It contains nine brief sections dealing with physician-to-physician relations and admonishing members to uphold the good name of the profession. "Every man who enters a fraternity," says the document, "engages by a tacit compact not only to submit to the law, but to promote the honor and interest of the association so far as is consistent with morality and the general good of mankind." But nowhere did the standing committee introduce the word ethics and the "tacit compact" mentioned was in no way comparable to the social contract of the political theorists of the Enlightenment, from Locke to Rousseau. Its closest analogue was the agreement between the medieval craftsman and his guild ([1], p. 86).

Chapman raises two intriguing questions: one historical, the other conceptual. The historical question is straightforward: was the Boston Medical Police a descendant of classical social contract theory? Chapman summarily dismisses this possibility, but the arguments in its favor are compelling. There is, in fact a direct line of textual descent from Locke's social contact to the passage Chapman cites in the Boston Medical Police. The passage was copied, word-for-word, from Chapter Two, Article XXIII, of Percival's *Medical Ethics*.

R. Baker (ed.), The Codification of Medical Morality, 25–39.
© 1995 *Kluwer Academic Publishers. Printed in the Netherlands.*

Percival, however, claims that he learned about "tacit compacts" from the Reverend Thomas Gisborne – to whom "a considerable portion of [the manuscript of *Medical Ethics*] was communicated" ([15], p. 5). Gisborne, in turn, had employed a theory of "tacit compact" in his *Enquiry into the Duties of Men in the Higher and Middle Classes of Society in Great Britain Resulting from Their Respective Stations, Professions and Employment* (1794, [7]) to justify attributing specific duties to various stations and offices, including "the professional office of physician." However, Gisborne had developed the theory of tacit compact in an earlier work, *The Principles of Moral Philosophy Investigated and Briefly Applied to the Constitution of Civil Society* (1789, [6]) to defend John Locke against William Paley's charge that social contracts were fictions, since no one had ever signed such a document. Gisborne's riposte was that signatures were irrelevant because those who accept society's protection and/or the privilege of its offices and stations, have *tacitly* committed themselves to a *compact*. The expression, "tacit compact", used by Percival in *Medical Ethics* and parroted in the Boston Medical Police, descends directly from terminology invented by Gisborne to defend Locke's social contract against Paley's critique – it is the language of the Enlightenment ideal of a social contract.

Chapman implicitly recognizes this when he notes that "the notion of [an] unwritten contract between patient and physician, and the primary emphasis on patient's rights, are unmistakable in Gisborne" but, he contends, they "were not carried over into Percival's [*Medical Ethics*] despite [Percival's] extravagant admiration for Gisborne's work" ([1], p. 83).

Why not? Why should Percival misrepresent his position? Chapman offers no explanation and the language and logic of Percival's text seem to support Percival's claim of indebtedness to Gisborne, rather than Chapman's denial. Gisborne's central terminological and conceptual innovations – the notion that duties of "professional office" are a "public trust" conferred by a "tacit compact" – permeate *Medical Ethics*. Here, for example, is Percival on the aging physician's duty to retire.

> The commencement of that period of senescence, when it becomes incumbent on the physician to decline the *offices of this profession* . . . is not easy to ascertain. . . . As age advances, therefore, a physician should . . . scrutinize impartially the state of his faculties; that he may determine . . . the precise degree in which he is qualified to exercise the active and multifarious *offices of his profession*. And whenever he becomes conscious that [his abilities have declined] . . . he should at once resolve, though others perceive not the changes which have taken place, to sacrifice every consideration of fame or fortune, and to retire from engagements of business. To the surgeon under similar circumstances, this rule is still more necessary. . . . Let both the physician

and surgeon never forget, that their *professions are public trusts*, properly rendered lucrative whilst they fulfil them; but which they are *bound*, by honor and probity, to relinquish, as soon as they find themselves unequal to their adequate and faithful execution ([15] Chapter Two, Article XXXII, *emphasis added*).

In this passage Percival draws on Gisborne's notion that as holders of "professional offices," physicians' privileges derive from a public trust. Consequently, when physicians find themselves "unequal to [the] adequate and faithful execution" of their end of the tacit compact, they are obligated "by honor and probity" to relinquish their professional office. The language and argument structure is patently Gisborne's and it is thus directly in the Enlightenment contractarian tradition.

Chapman could argue that even if, in drafting the Boston Medical Police, the committee parroted Percival's "tacit compact" language, and even if this language was understood by Percival to echo the Enlightenment ideal of a social contract, nonetheless, it does not follow that the committee drafting the Boston Medical Police intended anything ethical or enlightened by the language. Other historians have, in fact, raised questions about the extent to which the American codifiers understood Percival. John Haller observes that, "Curiously, American physicians showed little concern for the fact that Percival had written his code as an effort to sort out the competing interests of physicians, surgeons, and apothecaries in mill-town hospitals, such as the Manchester Infirmary, which he helped to organize – circumstances that were fundamentally different from American medical realities" ([10], p. 236). Moreover, as Chapman remarks, the nine brief sections of the Boston Medical Police deal entirely with intrapractitioner relationships; thus, unlike Percival, the Boston physicians seem to ignore physician-patient relationships. In what sense, then, can they be said to have established anything that might reasonable be called an Enlightenment social contract – or any other form of "medical ethics"?

This brings us to Chapman's conceptual question: Can a set of mutually advantageous arrangements between medical practitioners possibly constitute a medical ethic? On one analysis of ethics the answer is decidedly, "Yes." That analysis, of course, is contractarian.

The classic contractarians, Thomas Hobbes (1588–1679) and John Locke (1632–1704), appealed to the thought experiment of the social contract to reconstruct the rational core of morality, law, and government. The experiment was beguilingly simple. It turned on the question of what terms rational agents, not presently in a society, would accept in order to join one. The point of the thought-experiment was to discover both the terms that any rational agent would invariably *refuse* to accept as a condition for joining society, and those that they would *insist upon* in any society they joined. The former, they believed, constituted an agent's inalienable natural rights, the latter their fundamental civil rights. Hobbes believed that no rational agent would forego life to enter society, but he also argued that they would only enter a society if it accepted a civil morality of reciprocity (a version of the Golden Rule) and a civil law that treated all citizens equally. Locke expanded Hobbes' list of inalienable rights to incorporate liberty, health, and property, and more or less maintained Hobbes prerequisites of reciprocity and equality as the basis of civil morality and law.

Gisborne's *Principles of Moral Philosophy Investigated and Briefly Applied to the Constitution of Civil Society* (1789, [6]) is probably the first work of "applied" ethics in English; as its title proclaims, in the book Gisborne *applies* Locke's principles of equality and reciprocity to determine the moral 'constitution of civil society'." In that book, however, Gisborne worked only one side of Locke's analysis – the justification of liberty as an inalienable natural right – because he wanted to address the central issue of his day, the abolition of the slave trade. In his later book, *An Enquiry into the Duties of Men in the Higher and Middle Classes of Society in Great Britain Resulting from their Respective Stations, Professions and Employment* (1794), Gisborne worked the other side of the Lockean tradition, expanding Locke's principle of reciprocity to develop a tacit compact theory of professional obligation. Since it was only the later book that influenced Percival, it was this part of the contractarian tradition, the theory of reciprocity – not the natural rights theory that preoccupies Chapman – that Gisborne transmitted to Percival, and, through him, to the Boston physicians.

Contractarian theories based on the principle of reciprocity are still in the forefront of moral and political philosophy, including bioethics [3] , [21], thanks to the influence of three North American philosophers, David Gauthier [5], Robert Nozick [14] and John Rawls [16]. Gauthier, in particular, has developed a penetrating analysis of the role of the principle of reciprocity in classic contractarian thought. One of his more important insights is that cooperative paradoxes lie at the heart of the classic contractarianism. These paradoxes (for example the paradox of the commons, and the Prisoner's Dilemma) arise in contexts in which, even though non-cooperative behavior is ultimately disadvantageous to everyone, and, even though cooperative behavior is ultimately advantageous to everyone, at any given moment an individual can immediately better her or himself by betraying others (not reciprocating). Once betrayed, however, others tend to withdraw their cooperation, and – the heart of the dilemma – if no one cooperates, *everyone* is worse off, including the individual who originally chose to betray others to better her/himself. The irony of cooperative dilemmas is that although everyone loses, the ultimate cause of their loss is betrayal motivated by someone's desire to better themselves. Consequently, contractarians argue, everyone has an interest in *constraining* the urge to betray – perhaps especially those most tempted to betrayal. The classic mechanism of such constraint is the social contract.

To return to Chapman's question: Is the Boston Medical Police a social contractarian ethic? Since contractarian morality addresses dilemmas of social cooperation, the Boston Medical Police could only be a proper social contract if it addressed such a dilemma. As Chapman points out, the nine sections address physician-to-physician relationships – especially consultation. Thus the question of whether the Boston Medical Police is a social contract boils down to this: Did consultations pose a cooperative dilemma for nineteenth-century physicians? With rare unanimity, historians analyzing the period agree that it did.

The fractious state of the nineteenth-century American medical profession has often been remarked. William Rothstein illustrates the situation with a quotation from the *New York Monthly Chronicle of Medicine and Surgery* of 1825:

> No body of men are less in concert or seem less influenced by *esprit du corps*, than physicians. . . . The quarrels of physicians are proverbially frequent and bitter, intensity and duration seem to exceed that of other men. This state of things is in some degree attributable to the nature of the profession ([19], pp. 63–4, originally cited in [17], p. 2).

Hobbes himself never penned a more apt description of the war of each against all; of men unable to cooperate, reduced to unending quarrel, simply because they are caught in a cooperative dilemma. As sociologist Paul Starr observes, quoting Benjamin Rush, one of the primary source of these disputes was physicians inability to consult successfully.

> Nothing weakened the medical profession more than the bitter feuds and divisions that plagued doctors through the late nineteenth century. . . . They were open and acrimonious, and as common in the high tiers of the profession as in the low. Philadelphia, the center of early American medicine, was a maelstrom of professional ill will. . . . During the yellow fever epidemic in Philadelphia in 1793, Benjamin Rush and his rivals took to the press to denounce each other's treatment. "A Mahometan and a Jew," Rush wrote, "might as well attempt to worship the Supreme Being in the same temple, and through the medium of the same ceremonies, as two physicians of opposite principles and practice, attempt to confer about the life of the same patient." ([20], p. 93).

Why were "consultations" so problematic? Historian Charles Rosenberg reminds us, in a 1967 essay, "The Practice of Medicine in New York a Century Ago," that nineteenth-century consultations were fundamentally different from the contemporary variety.

> The emphasis upon the ethics of consultation and the frequency of such occasions emphasizes another quality of the profession a century ago. I refer to its still essentially "horizontal," undifferentiated structure. . . . Since almost all physicians did the same sort of things, consultants were ordinarily potential competitors. The general practitioner today sees, of course, little danger in referring cases demanding specialized knowledge to another physician; the functional differentiation of the profession within the past century has defined, and thus limited the specialist's activities ([18], pp. 223–253, reprinted in [12], pp. 58–59).

Thus, because of the horizontal structure of nineteenth-century medicine, when one physician consulted another, he was dealing with someone who had an immediate self-interest in disparaging his care of the patient. This placed physicians in a cooperative dilemma: for while, at any given time, they (and their patient) would be better off were a consultation to be arranged, to consult was always to place one's self at risk of betrayal.

Rothstein too emphasizes the dilemmatic nature of consultations as "a major source of conflict among physicians" ([19], p. 83), citing an essay by Daniel Drake (1785–1852), that was republished in 1832. Notice how in explaining the problems of consultation, Drake highlights the problem of betrayal.

> Consultations are copious sources of personal difficulty in the profession . . . Great reliance is, generally, placed by the patient . . . on the consulting physicians, because the other is presumed to have exhausted his skill. Should the patient die, it is often supposed that he might have lived had the consultation been held earlier. Thus the consulting physician, had nothing to lose, and much to gain. . . . The consulting physician, moreover, is often questioned, apart from the other [physician], on the past treatment and the probable issue of the case; when, if deficient in honor, he is apt to say, or look or insinuate, such things as he knows will operate to the injury of his colleague; who of course resents the insidious attack on his character should he discover it ([19] p. 83, quoting [4], p. 101).

Consultations, therefore, generated a classic dilemma of social cooperation: although physicians would be better off were everyone to practice consultation, any individual physician might do even better for himself by denigrating his colleagues during a consultation. Yet were any sizable number of physicians to denigrate their fellows, the practice of consultation would be untenable – and everyone would be worse off.

As every historian who has written on the subject properly observes, from the 1808 *Boston Medical Police*, to the 1847 *AMA Code of Ethics*, nineteenth-century American codes of ethics specifically addressed the dilemma of consultation. "Physicians," Rothstein writes, "viewed the code clauses involving consultations as being of particular importance . . . when the Medical Society . . . of New York developed its first code in 1823, the drafting committee was instructed to give special attention to consultations" ([19], p. 83). The nineteenth-century codes were thus mini-social contracts intentionally designed to resolve a cooperative dilemma. This explains why Percival's text provided such ideal source-material for the American codifiers; for, despite all the differences about the problems facing nineteenth- century American physicians and those Percival was dealing with at the Manchester Infirmary during the 1790s – which John Haller quite correctly remarks – fundamentally, they faced exactly the same problem, a cooperative paradox. Consequently, with a few minor modifications Percival's contractarian solution served American physicians in Boston as well as it had served British physicians and surgeons in Manchester.

How effective were these mini-social contracts at resolving the cooperative paradox of consultation? To analyze the actual function of these codes, Rothstein turns to the John Nichols and associates' detailed study of the Medical Society of the District of Columbia [13]. The first DC medical society was organized under an 1817 Congressional charter that permitted licensing, promoted science, but *prohibited* codes of ethics. Within three years "conflicts within the profession became so rife that the society became moribund and lost its charter" ([19], p. 80). In 1838, however, physicians founded an uncharted medical society that subscribed to a code of ethics. Although the two societies were to merge in 1911, it was the second society, the society with a code of ethics, that was "the more powerful and important in the District" ([19], p, 81).

Was the code of ethics integral to the success of the second medical society? Nichols believes that it was, remarking that the consultation provisions "came to acquire almost the force of a moral principle and a point of honourable conduct in the ideology of the members. . . . Lists of members were frequently issued, to show who were entitled to the privilege of consultations" ([13], Pt. II p. 29, quoted at [19], p. 83). History creates few controlled experiments, but the experience of the two in the District of Columbia comes close. The society that embraced an ethics code strongly regulating consultations, and that expelled physicians who refused to engage in reciprocal cooperation, flourished; the other society, prohibited by Congress from establishing a code of ethics and thus unable to constrain non-cooperation, succumbed to a Hobbesean war of each against all and became moribund.

But, one might query, did it really matter to nineteenth century physicians whether they were considered "entitled to the privilege of consultations"? After a comprehensive review of Massachusetts cases, Joseph Kett concluded that "the police power of the society, specifically the threat of expulsion, was adequate to bring many recalcitrants into line." ([11], p. 25). One revealing test of this power (to return to Boston and its *Medical Police*) involved Joseph Stephen Bartlett. Bartlett received his medical doctorate from Harvard University in 1831. He immediately joined the Boston Medical Society; in 1833 he joined its affiliate the Massachusetts Medical Society. In 1836, he was charged with fee-splitting, publishing testimonials for secret remedies, and consulting with non-members and expelled from both societies. Bartlett freely admitted that he had consulted with Dr. John Williams (an itinerant English oculist with a medical degree from Paris) and published testimonials to Williams' cures. He also acknowledged that his association with Williams violated the pledges he had signed to comply with the codes of ethics of both the Boston and Massachusetts medical societies. But, he contended, he had signed these pledges for economic reasons and, as a licensed practitioner with a degree from Harvard, he had a right to consult with whomever he wished, and to prescribe whatever he believed best for his patients: "I told the society, I should persevere in violating the by-laws whenever I thought the good of mankind required. I would not violate my conscience or my religion" ([2], p. 19).

The Boston and Massachusetts Medical societies responded to Bartlett's brazen defiance of their code of ethics by expelling him and by prohibiting society members from consulting with him. Bartlett seemed only slightly concerned over his expulsion, but, as explained to a special committee of the Massachusetts legislature to whom he appealed, the prohibition on consultation had destroyed his livelihood.

> The effects and influence of my expulsion have been highly injurious to my character and prospects. Medical gentlemen have refused to consult with me, and I have lost prospective practice. My degree from Harvard has been of no service to me since my expulsion. I have no right which any loafer might not enjoy. My diploma is of no use to me here. The influence of the Medical Society is to crush a man down and render him worse than dead. . . . I have had to struggle to sustain myself. . . . I rely upon my professional services for support. I find it necessary to follow other pursuits for a livelihood ([2], p. 15).

Bartlett, like any other plaintiff, had reason to overstate his losses. Nonetheless, other trial testimony confirms that his expulsion from the Boston Medical Society forced him to move to Marblehead, and that, were the expulsion from the Massachusetts Medical Society allowed to stand, he might be driven out of the state. We thus have a clear indication of how important consultations were to nineteenth century medical practice – without them a physician's livelihood was in jeopardy.

The grounds on which Bartlett was expelled tell us a great deal about protections offered to patients by the Boston Medical Police. The testimony of Mrs. Abigail Plumer, an impoverished blind woman, is particularly revealing. A neighbor had read Bartlett's published accounts of Williams's cures to her, and so Abigail traveled to Boston to put herself under Dr. Williams' care. Until this time Abigail's physician had been a Dr. Peirson, "who" as was expected of a medical society member, "charged me nothing" ([2], p. 32), as did her other physician, Dr. Reynolds (also a member of the Massachusetts Medical Society). Dr. Williams had different expectations.

When I first went to Dr. Williams he inquired how much money I had, and whether the people I lived with were rich or poor? . . . He said . . . that the medicine he gave me would restore my sight. I asked him his charge. He said I must pay $ 50 down, and $ 50 when cured. He said he could cure me. I did not feel able to pay the sum, and asked him if he would not take $ 25 down. He replied, [if] I set more by my money than my sight, he would have nothing to do with me. I told him I had but $ 43; that I had a child to support, and could lay by but little. He said he would not take less than $ 50, and I might have a week to make up my mind. I went home and returned the next Saturday with the money, which I gave him. After I paid him the money, he gave me some medicine in a bottle, and I signed a paper promising to pay him $ 50 more, if cured. The paper was read to me and [a witness] signed it with me. I then went back to Salem. . . . Williams charged me not to let any physician see the medicine. He told me that in six weeks I should be restored to sight. I went up to see [Williams] six times. . . . On my fourth visit Williams introduce[d] me to his friend Dr. Bartlett who would tell me of the cures [Williams] had effected. Dr. Bartlett examined my eyes; he said that mine was a case he would not like to undertake himself, but if any one could cure me it was Williams. This gave me courage. There was a woman in the room called Hannah. Dr. Bartlett said a cataract was coming off her eyes, and would be off in two or three days. I do not know that Dr. Bartlett had any connection with Williams. . . . At the end of two months, Williams said my eyes were doing well, and he should expect the other $ 50 sent on to him at Providence. He told me the cataracts are growing thinner, though I could not see the light of a candle at the time. . . . Dr. Bartlett said if I had put myself under Williams's care at the time I first lost my sight, it might have been saved. . . . [By contrast] Dr. Reynolds told me that had I come to him before he could have done nothing more than Peirson had done.

Dr. Peirson aided me in getting into an asylum. Benevolent friends in Salem procured me a loom – I learnt to weave mats and can now earn something for my support ([2], pp. 32–3).

Abigail Plumer's account provides damning evidence that Williams' practices violated the ethical codes of the two medical societies, and that Bartlett, by knowingly abetting these practices, violated them as well. Whereas Drs. Peirson and Reynolds abided by medical society rules, refusing to denigrate each other in consultation, or to charge Abigail Plumer, Bartlett disparaged other physicians and Williams set high fees for Abigail Plumer, even though she was poor. Even more damning, Williams was selling a secret nostrum. Bartlett admitted to this in direct testimony: Williams "keeps his remedial agents a secret, and he would be a great fool if he did not, until he had acquired an independent competency for himself and his family, as a reward for his labor" ([2], p. 13). Bartlett also acknowledged writing testimonials for Williams' secret remedy. Moreover, according to the testimony of Abigail Plumer, Bartlett gave private testimonials to Williams' patients – and if he were doing this for a fee, was fee-splitting.

Finally, Abigail Plumer's testimony suggests that Bartlett may have violated the ethics of consultation when he claimed that she might have been cured by going to Williams earlier. It is perhaps worth remarking that testimony given by Williams' other patients called this claim into question: all of them complained of extortionate fees but none were cured of cataracts or eye and ear ailments – however early or late they began their treatment. Indeed, when Bartlett himself was asked, under oath, "Have you seen such cases cured?" he replied:

> I saw one in New York greatly relieved by Dr. Williams. I saw the woman in September last; she then could discern nothing distinctly. I saw her a few weeks since, and she counted my fingers ([2], p. 3).

This one nameless out-of-state woman was all the direct evidence that Bartlett could adduce to substantiate the efficacy of Dr. Williams's secret remedy. He named no Massachusetts patients who could attest to a cure, and none came forward.

To return to an earlier point, Bartlett admitted his activities violated the *Boston Medical Police* and the Massachusetts Medical Society's code of ethics.

> There is a by-law of the Mass. Medical Society which makes it unlawful for a member to offer to cure disease by use of a secret medicine. He is bound to make known all his discoveries in medical science for the general good. There is no division of fees among the members of the society. Each receives and enjoys his own. ([2], p. 15)

Bartlett, however, rejects these constraints forthrightly:

> Harvard College has a right to confer certain privileges; that among these is the right of consultation, without referring to any body of men; that his right is inalienable, whatever may be the subsequent conduct of the individual upon whom it is conferred; . . . the Massachusetts

Medical Society contravenes [these rights] . . . [and] has imposed a yoke grievous to be borne, and that young physicians have suffered from its influence; [and that the Society violated his right of free speech] ([2], pp. 8–9).

Bartlett was thus claiming that a medical degree, like a driver's license, can be used in whatever lawful way, and for whatever lawful purposes, its holder wishes. No self-appointed society could constrain the lawful practices of the holder of a degree – including publishing testimonials, using secret remedies, consulting irregular practitioners, charging poor patients, or fee-splitting.

What became of Bartlett's challenge to the codes of medical ethics? Nothing. Bartlett died in the course of his trial and the Massachusetts legislature, having no petition before it, decided to let the powers of the Boston Medical Association and the Massachusetts Medical Society remain unchallenged. Nonetheless the question he posed remains: What justifies the special prerogatives of professional associations? The revisionist answer, Chapman's answer, is also Bartlett's: these so-called codes of ethics are merely guild rules, trade union regulations enacted to fill the purses of the membership. Yet the facts that came out in Bartlett's trial suggest a very different view: it was not the Medical Society members, Peirson and Reynolds, who sought to enrich themselves at the expense of a poor blind woman, it was the renegade Bartlett and the non-member Williams. *The Boston Medical Police*, in the section "Conduct for the Support of the Medical Character," echoes Benjamin Rush's position that: "Gratuitous services to the poor, are by no means prohibited; the characteristic beneficence of the profession, is inconsistent with . . . avaricious rapacity. The poor of every description should be especial objects of our peculiar care" ([22], p. 44). Peirson and Reynold's treated Abigail Plumer free of charge; Williams's treatment of her epitomizes avaricious rapacity.

Williams' avarice, however, did not provide the grounds for Bartlett's expulsion from the two medical societies. He was expelled for his public endorsement of Williams's secret nostrum. Secret nostrums were prohibited by both societies.

DISCOURAGEMENT OF QUACKERY

The use of quack medicines should be discouraged by the faculty, as disgraceful to the profession, injurious to health, and often destructive even of life. No physician or surgeon, therefore, should dispense a secret nostrum, whether [or not] it be his invention or exclusive property; for if it is of real efficacy the concealment of it is inconsistent with beneficence, and professional liberality; and if mystery alone give it value and importance, such craft implies either disgraceful ignorance, or fraudulent avarice ([22], p. 44).

This passage from the *Boston Medical Police* repeats word-for-word, Chapter Two, Articles XXI and XXII of Percival's *Medical Ethics*. It is one of the few places at which Percival places himself directly at odds with Professor John Gregory, whose *Lectures on the Duties and Qualifications of a Physician* [9] are one of Percival's acknowledged sources for *Medical Ethics*.

In the first of his lectures Gregory raises what is perhaps the must fundamental of all questions about medical ethics: Is "medicine [to] be considered . . . as an art, the most beneficial and important to mankind, or as a trade by which a considerable body of men gain their subsistence?" ([9], p. 13) Gregory believed medicine could be both: good medicine was good business – at least in the long run. It was bad business and "mean and selfish" for the medical faculty to refuse to prescribe those unorthodox remedies a patient expressly requested. Why should a trained practitioner lose business to quacks? Moreover, "every man has a right to speak where his life or his health is concerned" ([8], p. 31; [9], p. 35).

It is a physician's duty to do everything in his power that is not criminal, to save the life of his patient; and to search for remedies for every source, and from every hand, however mean and contemptible. This, it may be said, is sacrificing the dignity and interests of the faculty. But, I am not here speaking of the private police of a corporation, or the little arts of a craft. I am treating of the duties of a liberal profession, whose object is the life and health of the human species, a profession to be exercised by gentlemen of honour and ingenuous manners; the dignity of which can never be supported by means that are inconsistent with its ultimate object, and that tend only to increase the pride and fill the pockets of a few individuals ([8], p. 38; [9], p. 41).

Percival countered Gregory's liberal position on secret remedies by querying: why are the ingredients of secret remedies kept secret? If for reasons of "fraudulent avarice," then the rationale was clearly inconsistent with the beneficent goals of the profession. If, on the other hand, the remedy is effective, then concealing its ingredients also affronts professional beneficence, since secrecy prevents other physicians from prescribing the remedy to their patients. In either case, the purveyor of secret nostrums motives are suspect and are, moreover, inconsistent with tacit societal compact.

What drove the wedge between Gregory and Percival, here and elsewhere, was Gisborne's conception of the tacit compact. Gregory had envisioned medical morality as a virtue ethics, that is, he thought of morality in terms of the personal characteristics of practitioners. In his first lecture he lists "kinds of genius, understanding, and temper [that] naturally fit a man for being a physician" and "the moral qualities to be expected" from an individual "in the exercise of his profession, *viz.*, the obligations of humanity, patience, attention, discretion, secrecy and honour, which he lies under to his patients" ([9], pp. 11, 15). Thus morality, like genius was presumed to be individual. It is perhaps natural for Gregory, writing in the Edinburgh of Adam Smith, to presume a reductionist individualism and to dismiss all collective professional activity as mere "decorums . . . which tend most effectually to support the dignity of the profession; as likewise the propriety of his manners, his behavior to his patients, to his brethren, to surgeons and apothecaries."

Percival, following Gisborne, had reconceptualized all of the moral obligations listed by Gregory – humanity, attention, discretion, secrecy, and honor – in terms of a tacit compact between physicians and society. Collective professional activity thus became the basis of physicians' moral obligations. As a consequence, these obligations were more extensive than those envisioned by Gregory, expanding to encompass public health, concerns about aging physicians – and maintaining integrity and honor in consultations, refusing to prescribe secret remedies, refusing to split fees, and refraining from dealing with practitioners who refused to accept their professional obligations. Once Percival accepted Gisborne's analysis of medicine as a professional office conferred by tacit compact, he was forced to conclude that Gregory was wrong: good medicine – the duties of professional office – did not always coincide with good business. More specifically, one might have to loose business to quacks by refusing to prescribe secret nostrums.

We thus return to the passage from *the Boston Medical Police* cited by Chapman, the very passage that provided the grounds for expelling Bartlett from the Boston Medical Association, the "tacit compact" passage drawn from Chapter Two, Article XXII of Percival's *Medical Ethics* [15]:

CONDUCT FOR THE SUPPORT OF THE MEDICAL CHARACTER

The *esprit du corps* is a principle of action, founded in human nature, and, when duly regulated, is both rational and laudable. Every man, who enters into a fraternity, engages by a tacit compact, not only to submit to the laws, but to promote the honour and interest of the association, so far as they are consistent with morality and the general good of mankind. A physician, therefore, should cautiously guard against whatever my injure the general respectability of the profession, and should avoid all contumelious representations of the faculty at large, all charges against their selfishness or improbity, or the indulgence of an affected or jocular scepticism, concerning the efficacy and utility of the healing art ([22], p. 44).

In the context of the Boston Medical Police – and certainly for Joseph Stephen Bartlett – the five most important words in this passage are "to submit to the laws." Locke had held that people would naturally form voluntary associations based on the principle of reciprocity. Members of voluntary associations would naturally create laws for their governance – the social contract – and had a right to expel any members who refused to abide by them. Thus, precisely because the Boston Medical Association conceived of itself as a voluntary association that was tacitly obligated to serve society, it claimed the right to expel members who did not comply with its laws.

Was the Boston Medical Police an Enlightenment social contract in the Lockean tradition? Although in most respects it was, in one respect, properly noted by Chapman, it was not: Classic social contracts regulate the interactions of everyone affected by them, but the *Boston Medical Police* only regulated members of the Boston Medical Association. Since the Medical Association was granted a charter, government might be said to be a party to the compact, but what of patients? Without the participation of patients, the *Boston Medical Police* – indeed, virtually all codes of medical ethics drafted prior to the AMA *Code of Ethics* – was as best a partial social contract. All this changed in 1847, however, when the newly formed American Medical Association endorsed an explicitly contractarian code of medical ethics that was predicated upon the obligations of physicians to each other, to society, and to their patients – and upon the reciprocal obligations of the patients and society to physicians. Thus the AMA code represents the culmination of the Lockean contractarian ideal, initially applied to medicine by Gisborne, widely disseminated by Percival, and brought to America in the first code of medical ethics for a modern medical association, the *Boston Medical Police*.

BIBLIOGRAPHY

1. Chapman, C.: 1984, *Physicians, Law, and Ethics*, New York University Press, New York.
2. Committee of the Massachusetts State Legislature: 1839, *Report of the Evidence in the Case of John Stephen Bartlett, M.D., versus The Mass. Medical Society, as given before a Committee of the Legislature, at the Session of 1839*, House Report No. 76, Dutton and Wentworth State Printers, Boston.
3. Daniels, N.: 1985, *Just Health Care*, Cambridge University Press, Cambridge, UK.
4. Drake, D.: 1832, *Practical Essays on Medical Education and the Medical Profession in the United States*, Roff & Young, Cincinnati, reissued, 1954, Johns Hopkins University Press, Baltimore.
5. Gauthier, D.: 1986, *Morals by Agreement*, Oxford University Press, Oxford.
6. Gisborne, T.: 1789, *Principles of Moral Philosophy Investigated and Briefly Applied to the Constitution of Civil Society*, B. and J. White, London.
7. Gisborne, T.: 1794, *An Enquiry into the Duties of Men in the Higher and Middle Classes of Society in Great Britain Resulting from their Respective Stations, Professions and Employment*, B. and J. White, London.

8. Gregory, J.: 1770, *Observation on the Duties and Offices of a Physician and on the Method of Prosecuting Enquries in Philosophy*, W. Straham and T. Cadell, London.
9. Gregory, J.: 1772, *Lectures on the Duties and Offices of a Physician and on the Method of Prosecuting Enquries in Philosophy*, W. Straham and T. Cadell, London.
10. Haller, J.: 1981, *American Medicine in Transition 1840–1910*, University of Illinois Press, Urbana, IL.
11. Kett, J.: 1968, *The Formation of the American Medical Profession: The Role of Institutions, 1760–1860*, Yale University Press, New Haven.
12. Leavitt, J. and Numbers, R.: 1978, *Sickness and Health in America: Readings in the History of Medicine and Public Health*, University of Wisconsin Press, Madison, WI.
13. Nichols, J. *et al.*: 1947, *History of the Medical Society of the District of Columbia*, Washington DC.
14. Nozick, R.: 1974, *Anarchy, State and Utopia*, Basic Books, New York.
15. Percival, T.: 1803, *Medical Ethics; Or, A Code of Institutes and Precepts, Adapted to the Professional Conduct of Physicians and Surgeons*, J. Johnson, London.
16. Rawls, J.: 1971, *A Theory of Justice*, Harvard University Press, Cambridge, MA.
17. Rosen, G.: 1946, *Fees and Fee Bills*, Johns Hopkins University Press, Baltimore.
18. Rosenberg, C.: 1967, *Bulletin of the History of Medicine* 41, 223–53; in [12], pp. 58–9.
19. Rothstein, W.: 1972, *American Physicians in the Nineteenth Century: From Sects to Science*, The Johns Hopkins University Press, Baltimore.
20. Starr, P.: 1982, *The Social Transformation of American Medicine*, Basic Books, New York.
21. Veatch, R.: 1981, *A Theory of Medical Ethics*, Basic Books, New York.
22. Warren, J., Hayward, L., and Fleet, J.: 1808, *The Boston Medical Police*, Association of Boston Physicians, Boston, this volume, pp. 41–46.

JOHN WARREN, LEMUEL HAYWARD AND JOHN FLEET

BOSTON MEDICAL POLICE

Boston Medical Association (1808)

The standing Committee of the Association of Boston Physicians for the year, commencing on the first Wednesday of March, 1807, having been instructed to propose a code of Medical Police, to be submitted to the consideration of the Association at their next annual meeting, beg leave to report:

1. That having examined the different publications of Gregory, Rush and Percival upon this subject, they first selected from them such articles, as seemed most applicable to the circumstances of the profession in this place.
2. That with these articles as a ground work, they have proceeded to form a short system of police, containing general principles for the government of this Association, by making such alterations, or additions to them, as they thought necessary for rendering them both practicable and useful.
3. That they have added such new articles, as they judged conducive to the general views of this Association, and adapted to the particular situation of medical practice in America.

The result of which is submitted in the form following:

CONSULTATIONS

Consultations should be encouraged in difficult and protracted cases, as they give rise to confidence, energy, and more enlarged views in practice. On such occasions, no rivalship or jealousy should be indulged; candour, justice and all due respect should be exercized towards the physician who first attended; and as he may be presumed to be best acquainted with the patient and his family, he should deliver all the medical directions as agreed upon. It should be the

R. Baker (ed.), The Codification of Medical Morality, 41–46.
© 1995 *Kluwer Academic Publishers. Printed in the Netherlands.*

province, however, of the senior consulting physician to propose the necessary questions to the sick.

The consulting physician is never to visit without the attending one, unless by the desire of the latter, or when, as in sudden emergency, he is not to be found. No discussion of the case should take place before the patient or his friends; and no prognostications should be delivered, which were not the result of previous deliberation and concurrence. Theoretical debates, indeed, should generally be avoided in consultation, as occasioning perplexity and loss of time; for there may be much diversity of opinion on speculative points, with perfect agreement on those modes of practice, which are founded, not on hypothesis, but on experience and observation. Physicians in consultation, whatever may be their private resentments or opinions of one another, should divest themselves of all partialities, and think of nothing but what will most effectually contribute to the relief of those under their care.

If a physician cannot lay his hand to his heart and say, that his mind is perfectly open to conviction, from whatever quarter it may come, he should in honour decline the consultation.

All discussions and debates in consultations, are to be held secret and confidential.

Many advantages may arise from two consulting together, who are men of candour, and have mutual confidence in each other's honour. A remedy may occur to one, which did not to another, and a physician may want resolution or a confidence in his own opinion, to prescribe a powerful, but precarious remedy, on which, however, the life of his patient may depend; in this case, a concurrent opinion, may fix his own. But when such mutual confidence is wanting, a consultation had better be declined, especially if there is reason to believe, that sentiments delivered with openness, are to be communicated abroad, or to the family concerned; and if, in consequence of this, either gentleman is to be made responsible for the event.

The utmost punctuality should be observed in consultation visits; and to avoid loss of time, it will be expedient to establish the space of fifteen minutes, as an allowance for delay, after which, the meeting might be considered as postponed for a new appointment.

INTERFERENCES

Medicine is a liberal profession; the practitioners are, or ought to be men of education; and their expectations of business and employment should be founded on their degree of qualification, not on artifice and insinuation. A certain undefinable species of assiduities and attentions, therefore, to families usually employing another, is to be considered as beneath the dignity of a regular practitioner, and as making a mere trade of a learned profession; and all officious interferences in cases of sickness in such families, evince a meanness of disposition, unbecoming the character of a physician or a gentleman. No meddling inquiries should be made concerning them, nor hints given relative to their nature and treatment, nor any selfish conduct pursued, that may, directly or indirectly, tend to weaken confidence in the physicians or surgeons, who have the care of them.

When a physician is called to a patient, who has been under the care of another gentleman of the faculty, before any examination of the case he should ascertain, whether that gentleman has discontinued his visits, and whether the patient considers himself as under his care, in which case, he is not to assume the charge of the patient, nor to give his advice (excepting in instances of sudden attacks), without a regular consultation; and if such previously attending gentleman has been dismissed, or has voluntarily relinquished the patient, his practice should be treated with candour, and justified so far as probity and truth will permit; for the want of success in the primary treatment of the disorder, is no impeachment of professional skill and knowledge.

It frequently happens, that a physician, in incidental communications with the patients of others, or with their friends, may have their cases stated to him in so direct a manner, as not to admit of his declining to pay attention to them. Under such circumstances, his observations should be delivered with the most delicate propriety and reserve. He should not interfere in the curative plans pursued; and should even recommend a steady adherence to them, if they appear to merit approbation.

DIFFERENCES OF PHYSICIANS

The differences of physicians, when they end in appeals to the publick, generally hurt the contending parties; but, what is of more consequence, they discredit the profession, and expose the faculty itself to contempt and ridicule. Whenever such differences occur, as may affect the honour and dignity of the profession, and cannot immediately be terminated, or do not come under the character of violation of the special rules of the association, according to the nature of the dispute; but, neither the subject matter of such references, nor the adjudication, should, if it can be avoided, be communicated to the publick, as they may be personally injurious to the individuals concerned, and can hardly fail to hurt the general credit of the faculty.

DISCOURAGEMENT OF QUACKERY

The use of quack medicines should be discouraged by the faculty, as disgraceful to the profession, injurious to health, and often destructive even of life. No physician or surgeon, therefore, should dispense a secret nostrum, whether it be his invention or exclusive property; for if it is of real efficacy the concealment of it is inconsistent with beneficence, and professional liberality; and, if mystery alone give it value and importance, such craft implies, either disgraceful ignorance, or fraudulent avarice.

CONDUCT FOR THE SUPPORT OF THE MEDICAL CHARACTER

The *esprit du corps* is a principle of action, founded in human nature, and, when duly regulated, is both rational and laudable. Every man, who enters into a fraternity, engages, by a tacit compact, not only to submit to the laws, but to promote the honour and interest of the association, so far as they are consistent with morality and the general good of mankind. A physician, therefore, should cautiously guard against whatever may injure the general respectability of the profession, and should avoid all contumelious representations of the faculty at large, all general charges against their selfishness or improbity, or the indulgence of an affected or jocular scepticism, concerning the efficacy and utility of the healing art.

FEES

General rules are adopted by the faculty in every town, relative to the pecuniary acknowledgements of their patients; and it should be deemed a point of honour to adhere to them; and every deviation from, or evasion of these rules, should be considered as meriting the indignation and contempt of the fraternity.

Gratuitous services to the poor, are by no means prohibited; the characteristical beneficence of the profession, is inconsistent with sordid views and avaricious rapacity. The poor of every description should be the objects of our peculiar care. Dr. Boerhaave used to say, they were his best patients, because God was their paymaster.

It is obvious also, that an average fee, as suited to the general rank of patients, must be inadequate compensation from the rich (who often require attendance not absolutely necessary), and yet too large to be expected from that class of citizens, who would feel a reluctance in calling for assistance, without making some decent and satisfactory remuneration.

EXEMPTION FROM CHARGES

The clergymen of the town, and all members of the medical profession, together with their families, should be attended gratuitously; but visits should not be obtruded officiously, as such civility may give rise to embarrassments, or interfere with that choice on which confidence depends.

But distant members of the faculty, when they request attendance, should be expected to defray the charges of travelling; and such of clergy from abroad, as are qualified by their fortunes or incomes, to make a reasonable remuneration for medical attendance, are not more privileged, than any other order of patients.

Omission to charge on account of the wealthy circumstances of the physician, are an injury to the profession, as it is defrauding, in a degree, the common funds for its support, when fees are dispensed with, which might justly be claimed.

VICARIOUS OFFICES

Whenever a physician officiates for another by his desire, in consequence of sickness or absence, if for a short time only, the attendance should be performed gratuitously as to the physician, and with the utmost delicacy towards the professional character of the gentleman previously connected with the patient.

SENIORITY

A regular and academical education furnishes the only presumptive evidence of professional ability, and is so honourable and beneficial, that it gives a just claim to pre-eminence among physicians at large, in proportion to the degree in which it may be enjoyed and improved. Nevertheless, as industry and talents may furnish exceptions to this general rule, and this method may be liable to difficulties, in the application, seniority, among practitioners of this town, should be determined by the period of publick and acknowledged practice as a physician or surgeon in the same. This arrangement being clear and obvious, is adapted to remove all grounds of dispute amongst medical gentlemen; and it secures the regular continuance of the established order of precedency, which might otherwise be subject to troublesome interruptions, by new settlers, perhaps not long stationary in the place.

At a meeting of the Boston Medical Association, held at Vila's on the first Wednesday in March, 1808, the Committee of the preceding year, having, in conformity with their instructions, reported on a code of Medical Police, which was read and accepted by sections, it was voted,

That the Report of the Committee be recommitted, with instructions to print five hundred copies of the same and that they present to each member of the Association three copies of the Report, and distribute the remaining copies to such other Physicians of the State as they may think proper.

Voted likewise, that the thanks of the association be presented to the Committee for their judicious and useful Report.

J. Gorham, Secretary

CHAPTER 2

ROBERT BAKER

THE HISTORICAL CONTEXT OF THE AMERICAN MEDICAL ASSOCIATION'S 1847 *CODE OF ETHICS*

On the morning of May 7th, 1847, the national medical convention – soon to rename itself the American Medical Association – enacted a Code of Ethics. In the brief span of three years the frustration of a few New York physicians had led to the first national medical convention, to the first national medical association, and to the first national code of medical ethics. The object of the New Yorkers' frustration, however, was neither the lack of a national organization, nor the state of medical morals, it was the state of medical education. Bad education was driving out good: the shorter and cheaper the route to medical qualification, the fewer the demands a college made on a medical student, the more likely it was to be a financial success. Worse yet, piecemeal reform seemed impossible. No single state could reform its system without a parallel reform in the others. Medical students, medical qualifications, and medical practices were all portable; were any single state to raise educational standards within its borders, it would merely create a competitive advantage for colleges beyond its borders. Moreover, once students had earned their qualifications, however poorly educated they might be, they could still practice anywhere. Consequently, were any state to attempt to raise educational standards on its own, it was likely undercut its medical colleges, without significantly raising the qualifications of its physicians. Educational reform would either have to be done by everyone, everywhere, or no one could successfully implement it anywhere.

Reflecting back, Dr. Nathan Smith Davis of the Northwestern University Medical School recalls how this Gresham's Law scenario led him to propose a national medical convention.

R. Baker (ed.), The Codification of Medical Morality, 47–63.
© 1995 *Kluwer Academic Publishers. Printed in the Netherlands.*

The college degree of M.D., being almost everywhere accepted as authority to practice without other examinations, the college that offered to confer it after attendance on the shortest annual courses of instruction and the lowest college fees could generally draw the largest class.

Under these conditions and tendencies the annual courses of medical college instruction were progressively shortened from six months, as required by the first colleges in Philadelphia and New York, prior to 1800, to sixteen weeks or less; all semblance of a requirement of suitable preliminary education was omitted; and before the middle of the century had been reached the number of medical colleges had increased from four to forty, and the annual aggregate number of medical graduates from fifteen to more than one thousand. By nominally studying medicine for three years, including the two annual repetition courses of medical college instruction of less than four months each, the student could obtain a diploma entitling him to practice, which was easier and more economical than to study with a preceptor four years and pass an examination by the censors of a County or State Society.

. . . At the annual meeting of the New York Medical Society in February, 1844, I, then a young delegate from the Broome County Medical Society, presented a series of resolutions "declaring in favor of the adoption of a fair standard of general education for students before commencing the study of medicine; of lengthening the annual course of medical college instruction to at least six months with the grading of the curriculum of the studies; and of having all examination for license to practice medicine conducted by State Boards, independent of the colleges." After a brief discussion, the resolutions were laid on the table until the next meeting of the society, and copies . . . [were] sent to the several County Medical Societies...and to the medical periodicals. At the next meeting, [in] 1845, the resolutions were taken from the table and during a free discussion it was urged with much force that the requirements of a fair standard of education before commencing medical studies, a longer annual college term with proper grading of the curriculum, and independent examinations for license to practice in New York State alone, would only cause the student to abandon her colleges for those of Pennsylvania or the New England States.

This caused the original mover of the resolutions [i.e., Davis] to offer the following preamble and resolutions:

Whereas, it is believed that a National Convention would be conducive to the elevation of the standard of medical education in the United States, and

Whereas, there is no mode of accomplishing so desirable an object without concert of action on the part of the medical societies, colleges, and institutions of all the states, therefore

Resolved, That the New York State Medical Society earnestly recommends a National Convention of delegates from medical societies and colleges in the whole Union, to convene in the City of New York on the first Tuesday in May, 1846, for the purpose of adopting some concerted action on the subject set forth in the foregoing preamble ([4], pp. 142–43).

Davis's proposal was well received. On May 5, 1846, one hundred and twenty-two delegates – the largest group of American physicians ever assembled – convened at the medical department of the University of the City of New York. They came from fourteen states and represented sixteen medical societies (half state, half county/municipal), twelve medical colleges or institutes, two hospitals, and one asylum. Unfortunately, although the group was large, it was not representative: a fifth of the delegates were from New York, most of the midwest was absent, and New Jersey and some other major eastern states were without representation. Not surprisingly, therefore, one of the first motions put before the Convention was dissolution: Dr. Bedford of the University of City of New York moved to dissolve the convention on the grounds that it was too unrepresentative to reform American medical education.

Whereas the call of the State Medical Society of New York for a National Medical Convention ... has failed in a representation from one half the United States, and from a majority of the Medical Colleges; and whereas the State Medical Society has emphatically stated that there is no mode of accomplishing the object of the Convention, without concert of action on the part of the Medical Societies, Colleges and Institutions of *all* the States, therefore, *Resolved*, that this Convention adjourn *sine die* ([9], p. 15).

Bedford's motion was defeated by a vote of 74 to 2; nonetheless, by the second day, May 6, it became clear to the assemblage that Dr. Bedford had a point: the convention was not representative enough to undertake a reform of medical education. Faced with the prospect of utter failure, the conventioneers cast about for some concrete accomplishment to justify their enterprise. They found it in the following six motions proposed by Dr. Isaac Hays of Philadelphia.

1st *Resolved*, That it is expedient for the Medical Profession of the United States, to institute a *National Medical Association*, for the protection of their interests, for the maintenance of their honour, and respectability, for the advancement of their knowledge, and the extension of their usefulness.

2nd. *Resolved*, That a committee of seven be appointed to report a plan of organization for such an Association, at a meeting to be held in Philadelphia, on the first Wednesday in May, 1847.

3rd. that a committee of seven address all "regularly organized Medical Societies, and chartered Medical Schools in the United States, setting forth the objects of the National Medical Association, and inviting them to send delegates. . . ."

4th. *Resolved*, That it is desirable that a uniform and elevated standard of requirement for the degree of M.D., should be adopted by all the Medical Schools in the United States, and that a Committee be appointed to report on this subject [at the national meetings].

5th. *Resolved*, That it is desirable that young men before being received as students of Medicine, should have acquired a suitable preliminary education; and that a Committee [to be appointed, et

6th. *Resolved*, That it is expedient that the Medical Profession in the United States should be governed by the same code of Medical Ethics, and that a Committee of Seven be appointed to report a code for that purpose, at a meeting to be held at Philadelphia, on the first Wednesday of May, 1847.

Hays's six resolutions were passed unanimously by the sixty-three remaining delegates, and for good reason: by at once incorporating and transcending

Davis's initial concern with education, the resolutions provided the Convention with a *raison d'être*. The first and second resolutions boldly envision a permanent national medical organization that could address, not only questions of medical education, but any issue whatsoever that would affect the interests, honor, reputation, or usefulness of the medical profession, or the advancement of medical knowledge. This was so self-evidently a fitting (even if initially unintended) *raison d'être* for the first national medical convention, that the first two resolutions passed without debate. So did the third resolution, which reaffirms the alliance of medical societies and medical colleges originally envisioned by Davis and his fellow New Yorkers, and which sets up a committee to address the problem of representativeness. Resolutions four and five were more controversial but, in the end, two committees were set up to address Davis's original concerns about medical and pre-medical education. Resolution six sets up a committee to formulate a national code of medical ethics – introducing a subject distinct from anything contemplated by Davis and his fellow New Yorkers. It too passed unanimously.

The national code of medical ethics was to be the crowning achievement of the two national medical conventions, and of the American Medical Association in its early years. Yet it was not initially a major concern of the New Yorkers, and it was not on the official agenda for the First National Medical Convention. It appears to be put on the agenda of the Second National Medical Convention entirely at the behest of Isaac Hays – although it was unanimously approved. The source of the approval was easy to appreciate: since the Boston Medical Society issued its Medical Police in 1808, it had become fashionable for many, but by no means all, municipal, county and state medical societies to issue codes of medical ethics. The physicians assembled at the National Convention probably thought it natural that a new national medical association should consider issuing a code of ethics. Yet the timing was odd; for it is anything but obvious that a committee ought to formulate a national code ethics for a nascent national organization *before* the constitution of that organization was even drafted.

If the timing was odd from the perspective of the Convention, if medical ethics had not been on the minds of the New Yorkers who organized the First National Medical Convention, it nonetheless made excellent sense to Isaac Hays to make ethics an integral part of the agenda he proposed for a Second Convention. In 1846, Hays had more than an academic interest in medical ethics; throughout the year he had been entangled in a nasty law suit that eventually ended up before the Pennsylvania supreme court. As editor of *Medical News*, Hays had published an article declaring a popular nostrum inefficacious. The purveyors of the nostrum brought suit, arguing that his journal's assessment of their product had caused them to suffer a financial loss. Hays lost the suit and, while he escaped with only a token fine, the issue of the suit was potentially momentous; for the question raised by the suit was fundamental to the nature of medicine: was medicine a profession or a trade?

In *Lectures on the Duties and Offices of a Physician* [7], John Gregory had argued that medicine could be both a profession and a business, and that good physicians would have flourishing businesses. In *Medical Ethics* [10], however, Thomas Percival had held that the two were incompatible: that while the practice of medicine ought properly to be lucrative, it was first and foremost a professional office, not a business; that holders of this office were under an obligation to society to use scientific knowledge to alleviate human suffering; and that this obligation transcended obligations to hospital trustees, to patrons, and even one's own need to make a living [1]. Hays, however, had just lost a law suit which denied these obligations to science and society; a suit that, by reducing medicine to the status of a marketplace activity, to a trade, seemed to challenge the very idea of a scientific medicine and a scientific pharmacopoeia.

The issue facing Hays was akin to that raised by Davis: if medicine was merely a trade, then, as in any other business, that cheapest route to licensing was the best; if it were merely a business, pill-pushers and nostrum peddlers had an equal entitlement to sell their wares – *caveat emptor*; however, if, as Percival held, medicine was a profession constrained by societal obligations to apply science to healing, then both Davis's medical school reforms – the fundamental idea that to practice medicine one needed to learn the sciences of anatomy, physiology, pathology and biochemistry – and Hays's belief in publishing the results of scientific tests of drugs, were not only reasonable, they were morally obligatory.

The issue cut deeper. Hays believed that science advances as much by the detection of falsehood, error, and inefficacy, as by the discovery of truths. In 1826, for example, he had published a paper, "The Forces by Which the Blood is Circulated," in the *Philadelphia Journal of the Medical and Physical Sciences*. The paper challenged a theory propounded by none other than, Xavier Bichat, the founder of modern clinical medicine. Bichat held that the blood in capillaries does not circulate; however, Hays had found evidence indicating that the revered founder of clinical medicine was in error. Placing a commitment to truth over his veneration of a great man, Hays published his findings. As both researcher and editor, Hays argued that scientific medicine could only advance through the vigorous examination of theory and the continual renunciation of error. Yet the Pennsylvania courts had ruled that medicine was a matter of commerce, not science, and, as such, errors could not be publicly challenged. Hays quite properly treated the ruling as a fundamental challenge to scientific medicine; the suit was thus very much on his personal agenda in 1846 and he was sufficiently influential in shaping the agenda for the Second National Medical Convention, to put the question of a code of ethics on its agenda as well.

Hays had moved the second convention to his hometown, Philadelphia, and had himself put on the organizing committee, and the Medical Ethics Committee – although, characteristically, not as chair. The honor of chairing the Medical Ethics Committee went to his fellow Philadelphian, Dr. John Bell; with the addition of a third Philadelphian, Gouvernor Emerson, the largest delegation on the Committee were Philadelphians. Also serving were Drs. W. W. Morris, of nearby Delaware; T. C. Dunn of Rhode Island (who, like Bell and Hays, had received his M.D. from the University of Pennsylvania); A. Darius Clark of New York, and Richard. D. Arnold of Georgia (another University of Pennsylvania alumnus).

When the delegates to the Second Nation Medical Convention assembled in the Hall of the Academy of Natural Sciences at 10:00 AM, Wednesday, May 5, 1847, they found Dr. Isaac Hays there to greet them, not as President, or as one of the two Vice Presidents – these honors had been bestowed on Drs. J. Knight of New Haven, John Bell of Philadelphia, Edward Delafield, of New York, respectively – but, with characteristic "diffidence," as Chairman of the Committee on Arrangements. The Committee had done its job quite well: two-hundred and sixty-eight delegates from twenty-two states had their credentials accepted by the Convention, over twice the number of delegates, representing nearly twice the number of states, that had been represented in the previous year. Non-representativeness was no longer an issue, and the convention addressed itself to other matters – especially education.

The Convention did not consider the report of the Medical Ethics Committee until Friday morning, May 7th. Discussion was delayed, in part, because it had adjourned early the previous day to allow interested conventioneers to accept an invitation from Dr. Thomas Kirkbride to visit the Pennsylvania Hospital for the Insane. The real reason for delay, however, was that so much time was consumed in debates over medical education – about whether, for example, to require that medical students "attend upon hospital practice." By contrast, there was scarcely any debate of the *Code of Medical Ethics*. Dr. L. P. Bush of Delaware moved that the Convention adopt the whole report, including an Introduction written by John Bell. The only question raised from the convention floor came in the form of an amendment to the *Code* proposed by Dr. John

Atlee, of Pennsylvania, who moved to amend the statement "In consultations, the physician in attendance should deliver his opinion first; and when there are several consulting they should deliver their opinions *in the order in which they have been called in*" (II, iv, 4) so that the italicized words would be replaced by the phrase "in the order of seniority, commencing with the youngest."

As Hays remarked in a brief note prefacing the *Code*, most of the its language was drawn from Percival's *Medical Ethics*. The line in question is drawn from Chapter I, Article XIX which states, "In consultations on medical cases the junior physician should *deliver* his *opinion* first, and the others in progressive order of their seniority." Thus Atlee's amendment would restore the system of inverse hierarchical reporting (juniors first, seniors last) that Percival recommends throughout *Medical Ethics*. Inverse hierarchical reporting is a scheme designed to promote a free flow of ideas and information by minimizing the dampening effects of hierarchy; that is, the natural disinclination of subordinates to take a public stand that contravene positions taken out by their "superiors." Hays had systematically striped out the passages that overtly recognize cast, class, and social hierarchy to make *Medical Ethics* – a code of ethics that Percival designed for the ultra-hierarchical caste- and class-conscious world of eighteenth-century England – palatable to the egalitarian cast of American thought [1]. Apparently Hays had a keen appreciation of the American temperament; Atlee's amendment was defeated. Doctor Bush's motion was then put to the convention and passed. The American Medical Association had a *Code of Ethics*.

The *Code* that passed that morning, and that was officially published on June 5, 1847, differs significantly from the various printings of the *Code* from 1848 to the present day. The document the Convention passed had three distinct parts: first, a long introduction, for which John Bell explicitly and proudly claims authorship; second, a short note by Isaac Hays in which he denies authorship and editorship of the *Code* – attributing the former to Thomas Percival and Benjamin Rush, and the latter to the committee as a whole –; and third, the *Code of Ethics* itself. The contrast between Bell's assertions of authorship and Hays' denials is striking. Both men received their doctorates from the University of Pennsylvania, both hold prestigious positions in Philadelphia medicine, but, whereas Hays effaces himself, speaking through the words of others, Bell, while stylishly appealing to the wisdom of Hippocrates and learned physicians throughout the ages, nonetheless speaks in his own voice. He is especially emphatic about the physician's duty to testify "against quackery in all its forms"; he forthrightly denounces the "anomaly in legislation and penal enactments," which while "stringent for the repression and punishment of fraud in general . . . are silent, and of course inoperative, in the cases of both fraud and poisoning so extensively carried on by the host of quacks who infest the land." He deplores the advertisement-hungry press for being "too ready for the sake of lucre to aid and abet the enormities of quackery," and lauds "honourable exceptions" who swim against the tide. Bell calls upon "physicians, when themselves free from all taint . . . to direct the intention of the editors and proprietors of newspapers, and of periodical works in general, to the moral bearings of the subject." "Physicians," Bell argues, "can best see the extent of the evil."

This is precisely Hays fight. Yet Hays never speaks – except though the voice of Bell, Percival, and Rush. Why? Why use the words of others rather than one's own? By mid-nineteenth century the venerable tradition of placing one's thoughts in words attributed to famous forebearers had long been extinct. Bell, while genuflecting at the alter of ancient wisdom, had no compunction about using his own words and claiming authorship of the Introduction. Hays, by contrast, suggests that he, or rather the Committee, "had no ambition for the honor of authorship" because it found that the various codes of ethics "were based on that of Dr. Percival" and so they decided to follow "a similar course"; but the Committee did not follow this course. They added lines from Dr. Benjamin Rush; just as importantly, whereas most American codes of ethics actually claimed to be inspired by *both* Gregory and Percival, Hays – who had once written on Gregory – never mentions Gregory's name. Moreover, as Hays admits, he and the Committee were not scrupulously faithful to Percival's text, they "chang[ed] a word, or even a part of a sentence . . . and there are but few sections which have not undergone some modification." As Atlee's amendment makes clear, these modifications were not minor grammatical changes, since they often changed the meaning of statements, sometimes – as in the statement Atlee attempted to amend – actually inverting Percival's meaning. Consequently, as Hays admits, "for the language of many [statements], and for the arrangement of the whole, the Committee must be held exclusively responsible."

If the Committee had made so many changes, why did it not, claim authorship of a code inspired by Percival? Every previous state, county, and municipal code-drafting committee, commencing with Drs. John Warren, Lemuel Hayward, and John Fleet, the proud authors of the Boston Medical Police, had laid claim to inspiration from Gregory, Percival and Rush *and* had claimed authorship of the code they had drafted – even though most of these codes simply mimed the language of their predecessors. Yet Hays's Committee – which had jettisoned Gregory and significantly altered Percival – went out of its way to deny authorship, insisting that it had so carefully "preserved the words of Percival" that he, and not the Committee, should be considered the rightful author of the code. Why?

The answer has little to do with the politics of code-drafting and everything to do with the idiosyncrasies of one particular drafter, Isaac Hays. In his *Memoir of Isaac Hays, M.D.*, Dr. Alfred Stillé, Hays's friend and colleague, remarks on the long history of Hays's "natural diffidence." Again and again he reports that Hays declined a professorship here, a committee chairmanship there, inevitably preferring to serve rather than to lead, to organize rather than to chair, and to edit rather than to author. This diffident strain, this proclivity for declining advantages, for avoiding the limelight, is puzzling given Hays evident accomplishments – until one reaches the end of the memoir. There Stillé reports that on April 12, 1879, the eighty-three-year-old Hays was found dead in his chair, apparently reading a medical journal.

> By birth a Hebrew, he through long life adhered to the ancient faith; but while fixed in his own views he was entirely liberal to those of others, often quoting Pope's lines:
>
> For modes of faith let graceless zealots fight:
> His can't be wrong whose life is in the right.
>
> ([11], pp. 35–6).

One can't help but believe that it was the fear of provoking "graceless zealots" that led Hays, a practicing Jew in gentile America, not only to habitual diffidence but also to prefer to speak through the words of distinguished deceased gentiles. Bell took little risk in claiming authorship of his ideas, but not so with Hays; it was far safer and probably more effective to deny authorship and place it in the hands of that eminently Christian gentleman, Thomas Percival, then to face the disapprobation of "graceless zealots."

Hays's determination, for personal reasons, literally to write a code using Percival's words, created formidable problems. The most commonly parroted sections of *Medical Ethics* were excerpted from Chapter Two, dealing with private practitioner consultations and disputes (i.e., the material appropriated by the authors of the Boston Medical Police). In America, as in Britain, private practitioners competed with each other for paying patients and were thus continually tempted to disparage their competitor's character, skill and qualifications – even though they themselves recognized that to yield to this temptation was to directly undercut the profession and, indirectly, their own livelihoods. In parroting Percival, codes like the Boston Medical Police used his words to constrain reputation-bashing and to minimize the causes of intra-practitioner squabbles. As long as they quoted only Chapter Two – which dealt with private practitioners and private pay patients – the appropriation of Percival was relatively unproblematic.

Hays, however, also appropriated material from Chapter One, which dealt with the problems of hospital practitioners – university-educated physicians, hospital-trained surgeons, and apprenticeship-trained apothecaries – and with the treatment of the sick poor. Although Percival's intent was to diminish the significance of the medical status-hierarchy and to argue that the sick-poor in hospitals deserve the same treatment accorded to more affluent classes (see, Chapter One, Article II), his arguments and prose inevitably reflected the very distinctions of class, cast, and status that he was attempting to ameliorate. However, in America of 1847 – just a decade after Andrew Jackson completed his term as the seventh President of the United States – Jacksonian egalitarian sentiments were so deeply rooted, that any mention of the distinctions that Percival presumed, however well intended, was unacceptable; to put Hays's problem in contemporary terms – Percival's language in Chapter One was not "politically correct."

If text from Chapter One was to be appropriated, therefore, it had to be expurgated. Thus, whenever Percival refers to the classical medical hierarchy of "physician," "surgeon," and "apothecary," Hays excised these terms and the correlative distinctions, and replaced them with a single word, "physician." Where Percival constantly refers to different categories of practice and patients: private, hospital, dispensary, infirmary, lock hospital, and insane asylum, Hays reduced these categories with just two words, "hospital" and "patient." It was precisely the systematic expurgation of class and hierarchy from Chapter I, Article XIX, of Percival's text that Atlee's amendment challenged. For good Jacksonian reasons, Atlee had wanted to privilege subordinates over their nominal superiors; for even better Jacksonian reasons Hays expurgated the distinction altogether – and the Convention voted to uphold Hays's rewording.

Hays's radical expurgation of class, cast, and hierarchy, however, created the "arrangement" problem he refers to in his note to the Convention. In *Medical Ethics* Percival states the various obligations hospital physicians and surgeons have towards different types of charity hospital patients (Chapter One), and the obligations private physicians had towards their paying patients (Chapter Two). He enumerates a slightly different set of duties for each class of patients. Thus Chapter One, Article One reads:

I. HOSPITAL PHYSICIANS and SURGEONS should minister to the sick, with due impressions of the importance of their office. . . . They should study, also, in their deportment, so to unite *tenderness* with *steadiness*, and *condescension* with *authority*, as to inspire the minds of their patients with gratitude, respect, and confidence.

By contrast, Chapter Two, Article One reads:

I. The *moral rules of conduct*, prescribed towards hospital patients should be fully adopted in private or general practice. Every case, committed to the charge of a physician or surgeon, should be treated with attention, steadiness and humanity: Reasonable indulgences should be granted to the mental imbecilities and caprices of the sick: Secrecy and delicacy when required should be strictly observed.

What is striking is that even though the rules are said to be the same, patients in charity hospitals are to be treated with *tenderness, steadiness, condescension and authority*, whereas the self-paying private practice patient is to be treated with *attention, steadiness and humanity*. Hays had expunged the distinction between the two classes of patients and consequently he faced the problem of stating which of these duties apply. A perusal of Chapter I, Article I, Sections 1 and 2, of the AMA *Code of Ethics* reveals his solution: Hays imposes *all* these duties, on *all* physicians, towards *all* of their patients.

Hays's editing amounted to much more than egalitarian excisions and concomitant textual rearrangements. He inserts significant amounts of entirely new material and, just as importantly, some of his very small alterations of Percival's texts – a title here, a "therefore" there – fundamentally restructure Percival's basic argument. For Percival, the rules of conduct expounded in *Medical Ethics* were duties of "office," incurred by physicians because they have accepted privileges of practice conferred upon them by society, acting through government. As Percival himself admits ([10], p. 6) he borrowed this theoretical framework from the Rev. Thomas Gisborne's, *An Enquiry into the Duties of Men in the Higher and Middle Classes of Society in Great Britain Resulting from Their Respective Stations, Professions and Employment* (1794, [6]). Gisborne had held that those who accept from society the privileges of office, or of high station – for example, physicians, on the one hand, and gentleman, on the other – have thereby tacitly contracted to accept a set of duties to society. To enjoy position or privilege was automatically to incur social obligations.

The idea that physicians occupy a socially-conferred office seemed natural to Gisborne and Percival, both British subjects, who accepted it as a matter of historical fact that the privilege of practice had actually been bestowed upon British medical practitioners by the Crown and Parliament, for example, through the grant of Henry VIII to the Barber-Surgeons Guild (1512) and thorough the charter Royal College of the Physicians of London (1518). American medicine was essentially private, however, and was practiced without looking to government for permission. The very idea that medical practitioners were indebted to society, or to government, for bestowing upon them the privilege of practice made little or no sense – particularly at the founding convention of the American Medical Association. Thus Hays's decision to use Percival's words – especially those from Chapter One – was fundamentally problematic; the Gisbornean idea of physicians holding a societally-conferred office made little or no sense in the American context.

To adopt Percival to American culture, therefore, Hays had to provide an alternative philosophical framework for the duties propounded in the *Code*. Given his "diffidence," Hays had to find another Christian gentleman whose words could be arranged to support this framework. Dr. Benjamin Rush (1746–1813) a conveniently deceased fellow Philadelphian, an eminent Christian gentleman, a signer of the Declaration of Independence, and an authority often cited (but rarely used) by earlier code-authors filled the bill exactly. In "Duties of a Physician," an essay that, by coincidence, was published in 1794, the same year in which Gisborne published his *Enquiry*, Rush had suggested that the rights and duties of physicians and patients were reciprocal. The Medico-Chirurgical Society of Baltimore had appropriate Rush's words in their 1832 Code of Ethics; Bell and Hays followed their precedent, but used Rush's words differently, they used them to adapt Percival's justificatory framework to the American context – that is, they amalgamated Rush with Percival to develop the foundational idea of a tripartite social contract between physicians, patients, and society, in which each of the three parties has reciprocal obligations and rights.

Bell straightforwardly appropriates Rush's ideas in his Introduction to the *Code*; Hays characteristically makes the transformation more subtly, insinuating a "therefore" here, changing a chapter heading there. For example, Percival, who had envisioned his enterprise as simply enumerating the obligations of office, entitles his four chapters "Of Professional Conduct Relative to . . ." and then fills in the blanks with "Hospitals," "Private Practice," "Apothecaries," and, in Chapter Four, "certain cases which require a knowledge of the law." By contrast, each of the three chapters of Hays's *Code* refers to reciprocal obligations, *viz.*: "Chapter I. Of the Duties of Physicians to Their Patients, and of the Obligations of Patients to Their Physicians," "Chapter 2. Of the Duties of Physicians to Each Other, and to the Profession At Large," "Chapter 3. Of the Duties of the Profession to the Public, and of the Public to the Profession." The result, as Bell observes in his Introduction, is a new "medical deontology" that "compromise[s] not only the duties, but, also, the rights of a physician." The notion that physicians are to have rights as well as duties

– rights against their patients, rights against each other, and against society – is unlike anything in Percival or Gisborne. Percival, following Gisborne, had argued that physicians had duties; he had never suggested that they also had *rights*.

What rights do Bell and Hays ascribe to physicians in their social contract? Or, to ask the correlative question, under what obligations do they place others? Not surprisingly, given the emphasis on education in the medical conventions of 1846 and 1847, "The first duty of a patient is, to select as his medical adviser one who has received a regular professional education" (Chapter One, Article 2, Section 1). The other duties are elegantly spelled out in Bell's Introduction:

> Every duty or obligation implies, both in equity and for its successful discharge, a corresponding right. As it is the duty of a physician to advise, so has he an right to be attentively and respectfully listened to. Being required to expose his health and life for the benefit of the community, he has a just claim, in return, on all its members, collectively and individually, for aid to carry out his measures, and for all possible tenderness and regard to prevent needlessly harassing calls on his services and unnecessary exhaustion of his benevolent sympathies ([3], p. 66).

If Bell and Hays argued that physicians have rights, they were equally liberal in recognizing the reciprocal rights of patients and communities – rights that are much stronger than those imputed by Percival. At one point, Bell even suggests that patients have an inalienable right to health care – irrespective of their ability to pay.

> In thus deducing the rights of a physician from his duties, it is not meant to insist on such a correlative obligations, the withholding of the right exonerates from the discharge of the duty. Short of the formal abandonment of the practice of his profession, no medical man can withhold his services from the requisition either of an individual or a community, unless under the circumstances, of rare occurrence, in which his compliance would be not only unjust but degrading to himself, or to a professional brother, and so far diminish his future usefulness ([3], p. 67).

Hays, however, is more circumspect about physicians' obligations to deliver health care. In Chapter Three, Article One, Section 3, he argues that while "poverty, professional brotherhood," and the like "should always be recognized as presenting valid claims for gratuitous service," nonetheless "justice requires that some limits be placed on such good offices." In general, Hays prefers that eleemosynary services be rendered through public institutions or private charities and be partially compensated. At no point, however, may the pursuit of compensation permit a physician to abandon a patient, especially if the patient is incurable, since the patient's right to treatment is "far superior to all pecuniary considerations" (Chapter One, Article I, Section 5). Similarly, physicians may not abandon a community "when pestilence prevails, [for] it is their duty to face the danger, and to continue their labors for the alleviation of suffering, even at the jeopardy of their lives" (Chapter Three, Article I, Section 1).

Patients and communities not only have stronger rights in the *Code* than in *Medical Ethics*, patients also have rights of "secrecy" (i.e., confidentiality) that were inconceivable to Percival and Gisborne. Percival argues (Chapter Four, Article XIX) that practitioners should not be misled by "false tenderness or misguided conscience" into "withholding the necessary proofs" when testifying in criminal court. On the contrary, the practitioner is obligated "not to conceal any part of what he knows, whether interrogated particularly to that point or not." The priority accorded society and government here is a natural corollary of the Gisborne-Percival theory that physicians' obligations are really obligations to society and government, deriving from the government's granting of the privilege of practice. Bell and Hays, in contrast, believed that physicians' obligations arise from a direct tripartite social contract between physicians – society – and *patients*; consequently Hays is very protective of patients' secrets, arguing that patients have a strong right to confidentiality.

Secrecy and delicacy, when required by peculiar circumstances, should be strictly observed; and the familiar and confidential intercourse to which physicians are admitted in their professional visits, should be used with discretion, and with the most scrupulous regard to fidelity and honor. The obligation of secrecy extends beyond the period of professional services – none of the privacies of personal and domestic life, no infirmity of disposition or flaw of character observed during professional attendance, should ever be divulged by him except when he is imperatively required to do so. The force and necessity of this obligation are indeed so great, that professional men have, under certain circumstances, been protected in their observance of secrecy by courts of justice (Chapter I, Article I, Section 2).

In exchange for the stringent obligations Bell and Hays asked the American medical profession to impose upon itself, they believed that patients and the public ought to reciprocate by accepting certain responsibilities: foremost amongst these, as we noted earlier, was to distinguish between properly qualified, morally committed practitioners and quacks. Bell makes the argument as follows:

Medical ethics cannot be so divided as that one part shall obtain the full and proper force of moral obligations on physicians universally, and, at the same time, the other be construed in such a way as to free society from all restrictions in its conduct to them; leaving it to the caprice of the hour to determine whether the truly learned shall be overlooked in favor of ignorant pretenders – persons destitute alike of original talent and acquired fitness.

The choice is not indifferent, in an ethical point of view, besides its important bearing on the fate of the sick themselves, between the directness and sincerity of purpose, the honest zeal, the learning and impartial observances, accumulated from age to age for thousands of years, of the regularly initiated members of the medical profession, and the crooked devices and low arts, for evidently selfish ends, the unsupported promises and reckless trials of interloping empirics, whose very announcements of the means by which they profess to perform their wonders are, for most part, misleading and false, and, so far, fraudulent ([3], p. 66–67).

There is a parallel passage in Chapter Three, Article II, Section I of the *Code*, which contains a veiled but telling reference to Hays's personal experiences with the law.

The benefits accruing to the public directly and indirectly from the active and unwearied beneficence of the profession, are so numerous and important, that physicians are justly entitled to the utmost consideration and respect from the community. The public ought likewise to entertain a just appreciation of medical qualifications; – to make a proper discrimination between true science and the assumption of ignorance and empiricism, to afford every encouragement and facility for the acquisition of medical education – and no longer to allow the statute books to exhibit the anomaly of exacting knowledge from physicians, under liability to heavy penalties, and of making them obnoxious to punishment for resorting to the only means of obtaining it.

The incessant and sometimes strident denunciations of poorly qualified unscientific quacks and their nostrums that permeate both Bell's Introduction and Hays's *Code* go well beyond anything in Percival, Gisborne, or, for that matter, Gregory. Percival, whose most important achievement was to use the intersubjectivity of the scientific model as a paradigm for the resolution of moral disputes, nonetheless hesitates to denounce the unscientific practice of medicine. Physicians themselves are to practice scientific medicine and are thus expressly forbidden to dispense a secret nostrum (Chapter Two, Article XXII) and enjoined to discourage patients from using quack medicines; nonetheless, "some indulgence seems to be required to a credulity that is insurmountable: And the patient should neither incur the displeasure of the physician, nor be entirely deserted by him" (Chapter Two, Article XXI). The Boston Medical Police simply echoes Percival's views in these matters. In the *Code*, however, ambivalent toleration, is systematically replaced by intolerance. Chapter three, Article I, Section 4, proclaims that:

It is the duty of physicians, who are frequent witnesses of the enormities committed by quackery, and the injury to health and even destruction of life caused by the use of quack medicines, to enlighten the public on these subjects, to expose the injuries sustained by the unwary from the devices and pretensions of artful empirics and impostors. Physicians ought to use all the influence which they may possess, as professors in Colleges of Pharmacy, and by exercising their option in regard to the shops to which their prescriptions shall be sent, to discourage druggists and apothecaries from vending quack or secret medicines, or from being in any way engaged in their manufacture and sale.

The most intolerant – and ultimately the most controversial – line in the *Code* states that "no one can be considered as a regular practitioner, or fit associate in consultation, whose practice is based on an exclusive dogma" (Chapter Two, Article IV, Section 1). In these few words the *Code* expressly prohibited consultation with non-orthodox practitioners. Percival, in contrast, is characteristically ambivalent about both the status of unorthodox practitioners and the value of medical education, holding that "A regular *academical education* the only presumptive evidence of professional ability," it is not "indispensably necessary to the attainment of knowledge, skill, and experience." Consequently, those without proper qualifications or training "should not be fastidiously excluded form the privileges of fellowship"; nor from consultations "as the good of the patient is the sole object in view...the aid of an intelligent practitioner ought to be received with candour and politeness and his advice adopted if agreeable to sound judgment" (Chapter Two, Article XI). The *Code*, however, is adamantly exclusionary, extending fellowship *only* to regular licensed practitioners who accept "the accumulated experience of the profession, and...the aids actually furnished by anatomy, physiology, pathology, and organic chemistry" (Chapter Two, Article IV, Section 1).

As Nathan Davis [4], Austin Flint [5], Worthington Hooker [8], and a host of other defenders of the *Code* were later to complain, these are but a few words in a long and complex document, but many readers reacted as if they constituted its entirety. They read these words as a declaration of war between the American Medical Association at war with every unconventional healer in America. In the end, however, Hays and his committee had little choice. As Bell eloquently explains in his Introduction:

Veracity, so requisite in all the relations of life, is a jewel of inestimable value in medical description and narrative, the luster of which ought never be tainted for a moment, by even the breath of suspicion. Physicians are peculiarly enjoined, by every consideration of honour and of conscientious regard for the health and lives of their fellow beings, not to advance any statement unsupported by positive facts, nor to hazard an opinion or hypothesis that is not the result of deliberate inquiry into all the data and bearings of which the subject is capable.

Hasty generalization, paradox and fanciful conjectures, repudiated at all times by sound logic, are open to the severest reprehension on the still higher grounds of humanity and morals. Their tendency and practical operation cannot fail to be eminently mischievous ([3], p. 70).

The founders of the American Medical Association and the drafters of its *Code* were united by a singular belief in science as the basis of medical education and practice; the *Code* they drafted and endorsed enshrined and defended an intersubjective, collaborative, consultative scientific method that would accept as truth *only* theories that survived empirical testing. How, then, could they possibly accept as equals or as associates those practitioners – Christian Scientists, Eclectics, homeopaths, Thomsonians, and the like – who refused to subject their theories to empirical scrutiny and scientific testing? As Bell observes, physicians who accept science as their paradigm were committed by their obligation to veracity – as well as by their humane and moral obligations to their patients – to refuse to validate a medicine predicated upon conjecture and hasty generalization. When the founders of the American Medical Association committed themselves to science as the basis of medical education and practice, they found themselves inextricably at war with all non-scientific practitioners and schools of practice. Chapter Two, Article IV, Section 1 may the official declaration of war, but the ensuing struggle between scientific and non-scientific medicine was inevitable, with or without a formal declaration of hostilities.

BIBLIOGRAPHY

1. Baker, R.: 1993, "Deciphering Percival's Code", in [2], pp. 179–212.
2. Baker, R., Porter, D., and Porter, R.: 1993, *The Codification of Medical Morality: Historical and Philosophical Studies of the Formalization of Western Medical Morality in the Eighteenth and Nineteenth Centuries: Volume One: Medical Ethics and Etiquette in the Eighteenth Century*, Kluwer Academic Publishers, Dordrecht.
3. Bell, J.: 1847, "Introduction," *Code of Ethics*, this volume, pp. 65–72.
4. Davis, N.: 1903, *History of Medicine, with the Code of Medical Ethics*, Cleveland Press, Chicago.
5. Flint, A.: 1895, *Medical Ethics and Etiquette: The Code of Ethics Adopted by the American Medical Association, with Commentaries*, D. Appleton, New York.
6. Gisborne, T.: 1794, *An Enquiry into the Duties of Men in the Higher and Middle Classes of Society in Great Britain Resulting from their Respective Stations, Professions and Employment*, B. and J. White, London.
7. Gregory, J.: 1772, *Lectures on the Duties and Offices of a Physician*, W. Straham and T. Cadell, London.
8. Hooker, W.: 1849, 1972, *Physician and Patient*, Arno Press (reprint), New York.
9. National Medical Convention: 1846, *Minutes of the Proceedings of the National Medical Convention, held in the City of New York, 1846.*
10. Percival, T.: 1803, *Medical Ethics; Or, A Code of Institutes and Precepts, Adapted to the Professional Conduct of Physicians and Surgeons*, J. Johnson, London.
11. Stillé, A.: 1880, *Memoir of Isaac Hays, M.D., Extracted from Transactions of the College of Physicians of Philadelphia* (3rd Series, Vol. V.), Philadelphia.

JOHN BELL

INTRODUCTION TO THE CODE OF MEDICAL ETHICS

Medical ethics, as a branch of general ethics, must rest on the basis of religion and morality. They comprise not only the duties, but, also, the rights of a physician: and, in this sense, they are identical with Medical Deontology – a term introduced by a late writer, who has taken the most comprehensive view of the subject [1].

In framing a code on this basis, we have the inestimable advantage of deducing its rules from the conduct of many eminent physicians who have adorned the profession by their learning and their piety. From the age of Hippocrates to the present time, the annals of every civilized people contain abundant evidences of the devotedness of medical men to the relief of their fellow creatures from pain and disease, regardless of the privation and danger, and not seldom obloquy, encountered in return; a sense of ethical obligation rising superior, in their minds, to considerations of personal advancement. Well and truly was it said by one of the most learned men of the last century: that the duties of a physicians were never more beautifully exemplified in the conduct of Hippocrates, nor more eloquently described in his writings.

We may here remark, that, if a state of probation be intended for moral discipline, there is, assuredly, much in the life of a physician to impart this salutary training, and to assure continuance in a course of self-denial, and, at the same time, of zealous and methodical efforts for the relief of the suffering and unfortunate, irrespective of rank or fortune, or fortuitous elevation of any kind.

A few considerations on the legitimate range of medical ethics will serve as an appropriate introduction to the requisite rules for our guidance in the complex relations of professional life.

R. Baker (ed.), The Codification of Medical Morality, 65–72.
© 1995 *Kluwer Academic Publishers. Printed in the Netherlands.*

Every duty or obligation implies, both in equity and for its successful discharge, a corresponding right. As it is the duty of a physician to advise, so has he a right to be attentively and respectfully listened to. Being required to expose his health and life for the benefit of the community, he has a just claim, in return, on all its members, collectively and individually, for aid to carry out his measures, and for all possible tenderness and regard to prevent needlessly harassing calls on his services and unnecessary exhaustion of his benevolent sympathies.

His zeal, talents, attainments and skill are qualities he holds in trust for the general good, and which cannot be prodigally spent, either through his own neglect or the inconsiderateness of others, without wrong and detriment to himself and to them.

The greater the importance of the subject and the more deeply interested all are in the issue, the more necessary is it that the physician – he who performs the chief part, and in whose judgment and discretion under Providence, life is secured and death turned aside – should be allowed the free use of his faculties, undisturbed by a querulous manner, and desponding, angry, or passionate interjections, under the plea of fear, or grief, or disappointment of cherished hopes, by the sick and their friends.

All persons privileged to enter the sick room, and the number ought to be very limited, are under equal obligations of reciprocal courtesy, kindness and respect; and, if any exception be admissible, it cannot be at the expense of the physician. His position, purposes and proper efforts eminently entitle him to, at least, the same respectful and considerate attentions that are paid, as a matter of course and apparently without constraint, to the clergyman, who is admitted to administer spiritual consolation, and to the lawyer, who comes to make the last will and testament.

Although professional duty requires of a physician, that he should have such a control over himself as not to betray strong emotion in the presence of his patient, nor to be thrown off his guard by the querulousness or even rudeness of the latter, or of his friends at the bedside, yet, and the fact ought to be generally known, many medical men, possessed of abundant attainments and resources, and so constitutionally timid and readily abashed as to lose much of their self-possession and usefulness at the critical moment, if opposition be abruptly interposed to any part of the plan which they are about devising for the benefit of their patients.

Medical ethics cannot be so divided as that one part shall obtain the full and proper force of moral obligations on physicians universally, and, at the same time, the other be construed in such a way as to free society from all restrictions in its conduct to them; leaving it to the caprice of the hour to determine whether the truly learned shall be overlooked in favor of ignorant pretenders – persons destitute alike of original talent and acquired fitness.

The choice is not indifferent, in an ethical point of view, besides its important bearing on the fate of the sick themselves, between the directness and sincerity of purpose, the honest zeal, the learning and impartial observances, accumulated from age to age for thousands of years, of the regularly initiated members of the medical profession, and the crooked devices and low arts, for evidently selfish ends, the unsupported promises and reckless trials of interloping empirics, whose very announcements of the means by which they profess to perform their wonders are, for most part, misleading and false, and, so far, fraudulent.

In thus deducing the rights of a physician from his duties, it is not meant to insist on such a correlative obligation, that the withholding of the right exonerates from the discharge of the duty. Short of the formal abandonment of the practice of his profession, no medical man can withhold his services from the requisition either of an individual or of the community, unless under the circumstances, of rare occurrence, in which his compliance would be not only unjust but degrading to himself, or to a professional brother, and so far diminish his future usefulness.

In the discharge of their duties to Society, physicians must be ever ready and prompt to administer professional aid to all applicants, without prior stipulation of personal advantages to themselves.

On them devolves, in a peculiar manner, the task of noting all the circumstances affecting the public health, and of displaying skill and ingenuity in devising the best means for its protection.

With them rests, also, the solemn duty of furnishing accurate medical testimony in all cases of criminal accusation of violence, by which health is endangered and life destroyed, and in those other numerous ones involving the question of mental sanity and of moral and legal responsibility.

On these subjects – Public Hygiene and Medical Jurisprudence – every medical man must be supposed to have prepared himself by study, observation, and the exercise of a sound judgment. They cannot be regarded in the light of accomplishments merely: they are an integral part of the science and practice of medicine.

It is a delicate and noble task, by the judicious application of Public Hygiene, to prevent disease and to prolong life; and thus to increase the productive industry, and, without assuming the office of moral and religious teaching, to add to the civilization of an entire people.

In the performance of this part of their duty, physicians are enabled to exhibit the close connection between hygiene and morals; since all the causes contributing to the former are nearly equally auxiliary to the latter.

Physicians, as conservators of the public health, are bound to bear emphatic testimony against quackery in all its forms; whether it appears with its usual effrontery, or masks itself under the garb of philanthropy and sometimes of religion itself.

By an anomaly in legislation and penal enactments, the laws, so stringent for the repression and punishment of fraud in general, and against attempts to sell poisonous substances for food, are silent, and of course inoperative, in the cases of both fraud and poisoning so extensively carried on by the host of quacks who infest the land.

The newspaper press, powerful in the correction of many abuses, is too ready for the sake of lucre to aid and abet the enormities of quackery Honourable exceptions to the once general practice in this respect are becoming, happily more numerous, and they might be more rapidly increased, if physicians, when themselves free from all taint, were to direct the intention of the editors and proprietors of newspapers, and of periodical works in general, to the moral bearings of the subject.

To those who, like physicians, can best see the extent of the evil, it is still more mortifying than in the instances already mentioned, to find members of other professions, and especially ministers of the Gospel, so prone to give their countenance, and, at times, direct patronage, to medical empirics, both by their use of nostrums, and by their certificates in favour of the absurd pretensions of the impostors.

The credulous, on these occasions, place themselves in the dilemma of bearing testimony either to a miracle or to an imposture: to a miracle, if one particular agent, and it often of known inertness or slight power, can cure all diseases, or even any one disease in all its stages; to an imposture, if the alleged cures are not made, as experiences shows that they are not.

But by no class are quack medicines and nostrums so largely sold and distributed as by apothecaries, whose position towards physicians, although it many not amount to actual affinity, is such that it ought, at least, to prevent them from entering into an actual, if not formally recognized, alliance with empirics of every grade and degree of pretention.

Too frequently we meet with physicians, who deem it a venial error, in ethics, to permit, and even to recommend, the use of quack medicine or secret compounds by their patients and friends. They forget that their toleration implies sanction of a recourse by the people generally to unknown, doubtful, and conjectural fashions of medication; and that the credulous in this way soon become the victims of an endless succession of empirics. It must have been generally noticed, also, that they whose faith is strongest in the most absurd pretensions of quackery, entertain the greatest skepticism towards regular and philosophic medicine.

Adverse alike to ethical propriety and to medical logic, are the various popular delusions which, like so many epidemics, have, in successive ages, excited the imagination with extravagant expectations of the cure of all diseases and the prolongation of life beyond its customary limits, by means of a single substances. Although it is not in the power of physicians to prevent, or always to arrest, these delusions in their progress, yet it is incumbent on them, from their superior knowledge and better opportunities, as well as from their elevated vocation, steadily to refuse to extend them the slightest countenance, still less support.

These delusions are sometimes manifested in the guise of a new and infallible system of medical practice, – the faith in which, among the excited believers, is usually in the inverse ratio of the amount of common sense evidence in its favour. Among the volunteer missionaries for its dissemination, it is painful to see members of the sacred profession, who, above all others, ought to keep aloof from vagaries of any description, and especially of those medical ones which are allied to empirical imposture.

The plea of good intention is not an adequate reason for the assumption of so grave a responsibility as the propagation of a theory and practice of medicine, of the real foundation and nature of which the mere medical amateur must necessarily, from his want of opportunities for study, observation, and careful comparison, be profoundly ignorant.

In their relations with the sick, physicians are bound, by every consideration of duty, to exercise the greatest kindness with the greatest circumspection; so that, whilst they make every allowance for impatience, irritation, and inconsistencies of manner and speech of the sufferers, and do their utmost to soothe and tranquilize, they shall, at the same time, elicit from them, and the persons in their confidence, a revelation of all the circumstance connected with the probable origin of the diseases which they are called upon to treat.

Owing either to the confusion and, at times, obliquity of mind produced by the disease, or to considerations of false delicacy and shame, the truth is not always directly reached on these occasions; and hence the necessity, on the part of the physicians, of a careful and minute investigation into both the physical and moral state of his patient.

A physician in attendance on a case should avoid expensive complication and tedious ceremonials, as being beneath the dignity of true science and embarrassing to the patient and his family, whose troubles are already great.

In their intercourse with each other, physicians will best consult and secure their own self-respect and consideration from society in general, by a uniform courtesy and high-minded conduct towards their professional brethren. The confidence in his intellectual and moral worth, which each member of the profession is ambitious of obtaining for himself among his associates, ought to make him willing to place the same confidence in the worth of others.

Veracity, so requisite in all the relations of life, is a jewel of inestimable value in medical description and narrative, the lustre of which ought never be tainted for a moment, by even the breath of suspicion. Physicians are peculiarly enjoined, by every consideration of honour and of conscientious regard for the health and lives of their fellow beings, not to advance any statement unsupported by positive facts, nor to hazard an opinion or hypothesis that is not the result of deliberate inquiry into all the data and bearings of which the subject is capable.

Hasty generalization, paradox and fanciful conjectures, repudiated at all times by sound logic, are open to the severest reprehension on the still higher grounds of humanity and morals. Their tendency and practical operation cannot fail to be eminently mischievous.

Among professional men associated together for the performance of professional duties in public institutions, such as Medical Colleges, Hospitals and Dispensaries, there ought to exist, not only harmonious intercourse, but also a general harmony in doctrines and practice; so that neither students nor patients shall be perplexed, nor the medical community mortified by contradictory views of the theory of disease, if not of the means of curing it.

The right of free inquiry, common to all, does not imply the utterance of crude hypotheses, the use of figurative language, a straining after novelty for novelty's sake, and the involution of old truths, for temporary effect and popularity, by medical writers and teachers. If, therefore, they who are engaged in a common cause, and for the furtherance of a common object, could make an offering of the extreme, the doubtful, and the redundant, at the shrine of philosophical truth, the general harmony in medical teaching, now desired, would be of easy attainment.

It is not enough, however, that the members of the medical profession be zealous, well informed and self-denying, unless the social principle be cultivated by their seeking frequent intercourse with each other, and cultivating, reciprocally, friendly habits of acting in common.

By union alone can medical men hope to sustain the dignity and extend the usefulness of their profession. Among the chief means to bring this desirable end, are frequent social meetings and regular organized Societies; a part of whose beneficial operation would be an agreement on a suitable standard of medical education, and a code of medical ethics.

Greatly increased influence, for the entire body of the profession, will be acquired by a union for the purposes of common benefit and the general good; while to its members, individually, will be insured a more pleasant and harmonious intercourse, one with another, and an avoidance of many heart burnings and jealousies, which originate in misconception, through misrepresentation on the part of individuals in general society, of each other's disposition, motives, and conduct.

In vain will physicians appeal to the intelligence and elevated feelings of the members of other professions, and of the better part of society in general, unless they be true to themselves, by a close adherence to their duties, and by firmly yet mildly insisting on their rights; and this not with a glimmering perception and faint avowal, but rather with a full understanding and firm conviction.

Impressed with the nobleness of their vocation, as trustees of science and almoners of benevolence and charity, physicians should use unceasing vigilance to prevent the introduction into their body of those who have not been prepared by a suitably preparatory moral and intellectual training.

No youth ought to be allowed to study medicine, whose capacity, good conduct, and elementary knowledge are not equal, at least, to the common standard of academic requirements.

Human life and human happiness must not be endangered by the incompetency of presumptuous pretenders,The greater the inherent difficulties of medicine, as a science, and the more numerous the complications that embarrass its practice, the more necessary is it that there should be minds of a higher order and thorough cultivation, to unravel its mysteries and to deduce scientific order from apparently empirical confusion.

We are under the strongest ethical obligations to preserve the character which has been awarded, by the most learned men and best judges of human nature, to the members of the medical profession, for general and extensive knowledge, great liberality and dignity of sentiment, and prompt effusions of beneficence.

In order that we may continue to merit these praises, every physician, within the circle of his acquaintance, should impress both fathers and sons with the range and variety of medical study, and with the necessity of those who desire to engage in it, possessing, not only good preliminary knowledge, but, likewise, some habits of regular and systematic thinking.

If able teachers and writers, and profound inquiries, be still called for to expound medical science, and to extend its domain of practical applications and usefulness, they cannot be procured by an intuitive effort on their own part, nor by the exercise of the elective suffrage on the part of others. They must be the product of a regular and comprehensive system, – members of a large class, from the great body of which they only differ by the course of fortuitous circumstances that gives them temporary vantage ground, for the display of qualities and attainments common to their brethren.

BIBLIOGRAPHY

1. Simon, M.I.: 1845, *Deontology Médicale, ou Des Devoirs et Des Droits Des Médecins Dans L'État Actuel De La Civilisation*, B. Bailliere, Paris.

JOHN BELL, ISAAC HAYS, G. EMERSON, W. W. MORRIS, T. C. DUNN, A. CLARK AND R. D. ARNOLD

NOTE TO CONVENTION

Doctor Hays on presenting this report stated that justice required some explanatory remarks should accompany it. The members of the Convention, he observed, would not fail to recognize in parts of it, expressions with which they were familiar. On examining a great number of codes of ethics adopted by different societies in the United States, it was found that they were all based on that by Dr. Percival, and that the phrases of this writer were preserved, to a considerable extent, in all of them. Believing that language that had been so often examined and adopted, must possess the greatest of merits for a document such as the present, clearness and precision, and having no ambition for the honours of authorship, the Committee which prepared this code have followed a similar course, and have carefully preserved the words of Percival wherever they convey the precepts it is wished to inculcate. A few of the sections are in the words of the late Dr. Rush, and one or two sentences are from other writers. But in all cases, wherever it was thought that the language could be made more explicit by changing a word, or even a part of a sentence, this has been unhesitatingly done; and thus there are but few sections which have not undergone some modification; while, for the language of many, and for the arrangement of the whole, the Committee must be held exclusively responsible.

SUBMISSION OF CODE OF ETHICS

The Committee appointed under the sixth resolution adopted by the Convention which assembled in New York in May last, to prepare a Code of Ethics for

R. Baker (ed.), The Codification of Medical Morality, 73–74.
© 1995 *Kluwer Academic Publishers. Printed in the Netherlands.*

the government of the medical profession of the United States respectfully submit the following Code. Philadelphia, June 5, 1847.

Committee
John Bell
Isaac Hays
G. Emerson
W. W. Morris
T. C. Dunn
A. Clark
R. D. Arnold

ISAAC HAYS

CODE OF ETHICS

CHAPTER I. OF THE DUTIES OF PHYSICIANS TO THEIR PATIENTS, AND OF THE OBLIGATIONS OF PATIENTS TO THEIR PHYSICIANS

Art. I – *Duties of Physicians to Their Patients*

1. A physician should not only be ever ready to obey the calls of the sick, but his mind ought also to be imbued with the greatness of his mission, and of the responsibility he habitually incurs in its discharge. Those obligations are the more deep and enduring, because there is no tribunal other than his own conscience, to adjudge penalties for carelessness or neglect. Physicians should, therefore, minister to the sick with due impressions of the importance of their office; reflecting that the ease, the health, and the lives of those committed to their charge, depend on their skill, attention and fidelity. They should study, also, in their deportment, so to unite tenderness with firmness, and condescension with authority, as to inspire the minds of their patients with gratitude, respect and confidence.
2. Every case committed to the charge of a physician should be treated with attention, steadiness and humanity. Reasonable indulgence should be granted to the mental imbecility and caprices of the sick. Secrecy and delicacy, when required by peculiar circumstances, should be strictly observed; and the familiar and confidential intercourse to which physicians are admitted in their professional visits, should be used with discretion, and with the most scrupulous regard to fidelity and honor. The obligation of secrecy extends beyond the period of professional services – none of the privacies of personal

R. Baker (ed.), The Codification of Medical Morality, 75–87.
© 1995 *Kluwer Academic Publishers. Printed in the Netherlands.*

and domestic life, no infirmity of disposition or flaw of character observed during professional attendance, should ever be divulged by him except when he is imperatively required to do so. The force and necessity of this obligation are indeed so great, that professional men have, under certain circumstances, been protected in their observance of secrecy by courts of justice.

3. Frequent visits to the sick are in general requisite, since they enable the physician to arrive at a more perfect knowledge of the disease – to meet promptly every change which may occur, and also tend to preserve the confidence of the patient. But unnecessary visits are to be avoided, as they give useless anxiety to the patient, tend to diminish the authority of the physician, and render him liable to be suspected of interested motives.
4. A physician should not be forward to make gloomy prognostications, because they savor of empiricism, by magnifying the importance of his services in the treatment or cure of the disease. But he should not fail, on proper occasions, to give to the friends of the patient timely notice of danger, when it really occurs; and even to the patient himself, if absolutely necessary. This office, however, is so peculiarly alarming when executed by him, that it ought to be declined whenever it can be assigned to any other person of sufficient judgment and delicacy. For, the physician should be the minister of hope and comfort to the sick; that, by such cordials to the drooping spirit, he may smooth the bed of death, revive expiring life, and counteract the depressing influence of those maladies which often disturb the tranquillity of the most resigned, in their last moments. The life of a sick person can be shortened not only by the acts, but also by the words or the manner of a physician. It is, therefore, a sacred duty to guard himself carefully in this respect, and to avoid all things which have a tendency to discourage the patient and to depress his spirits.
5. A physician ought not to abandon a patient because the case is deemed incurable; for his attendance may continue to be highly useful to the patient, and comforting to the relatives around him, even to the last period of a fatal malady, by alleviating pain and other symptoms, and by soothing mental anguish. To decline attendance, under such circumstances, would be sacrificing to fanciful delicacy and mistaken liberality, that moral duty, which is independent of, and far superior to all pecuniary consideration.
6. Consultations should be promoted in difficult or protracted cases, as they give rise to confidence, energy, and more enlarged views in practice.
7. The opportunity which a physician not unfrequently enjoys of promoting and strengthening the good resolutions of his patients, suffering under the consequences of vicious conduct, ought never to be neglected. His coun-

sels, or even remonstrances, will give satisfaction, not offense, if they be proffered with politeness, and evince a genuine love of virtue, accompanied by a sincere interest in the welfare of the person to whom they are addressed.

Art. II – *Obligations of Patients to Their Physicians*

1. The members of the medical profession, upon whom are enjoined the performance of so many important and arduous duties towards the community, and who are required to make so many sacrifices of comfort, ease, and health, for the welfare of those who avail themselves of their services, certainly have a right to expect and require, that their patients should entertain a just sense of the duties which they owe to their medical attendants.
2. The first duty of a patient is, to select as his medical adviser one who has received a regular professional education. In no trade or occupation do mankind rely on the skill of an untaught artist; and in medicine, confessedly the most difficult and intricate of the sciences, the world ought not to suppose that knowledge is intuitive.
3. Patients should prefer a physician whose habits of life are regular, and who is not devoted to company, pleasure, or to any pursuit incompatible with his professional obligations. A patient should also confide the care of himself and family, as much as possible, to one physician, for a medical man who has become acquainted with the peculiarities of constitution, habits, and predispositions, of those he attends, is more likely to be successful in his treatment than one who does not possess that knowledge.

   A patient who has thus selected his physician, should always apply for advice in whatever may appear to him trivial cases, for the most fatal results often supervene on the slightest accidents. It is of still more importance that he should apply for assistance in the forming stage of violent diseases; it is to a neglect of this precept that medicine owes much of the uncertainty and imperfection with which it has been reproached.
4. Patients should faithfully and unreservedly communicate to their physician the supposed cause of their disease. This is the more important, as many diseases of a mental origin simulate those depending on external causes, and yet are only to be cured by ministering to the mind diseased. A patient should never be afraid of thus making his physician his friend and adviser; he should always bear in mind that a medical man is under the strongest obligations of secrecy. Even the female sex should never allow feelings of shame and deli-

cacy to prevent their disclosing the seat, symptoms and causes of complaints peculiar to them. However commendable a modest reserve may be in the common occurrences of life, its strict observance in medicine is often attended with the most serious consequences, and a patient may sink under a painful and loathsome disease, which might have been readily prevented had timely intimation been given to the physician.

5. A patient should never weary his physician with a tedious detail of events or matters not appertaining to his disease. Even as relates to his actual symptoms, he will convey much more real information by giving clear answers to interrogatories, than by the most minute account of his own framing. Neither should he obtrude the details of his business nor the history of his family concerns.
6. The obedience of a patient to the prescriptions of his physician should be prompt and implicit. He should never permit his own crude opinions as to their fitness, to influence his attention to them. A failure in one particular may render an otherwise judicious treatment dangerous, and even fatal. This remark is equally applicable to diet, drink, and exercise. As patients become convalescent, they are very apt to suppose that the rules prescribed for them may be disregarded, and the consequence, but too often, is a relapse. Patients should never allow themselves to be persuaded to take any medicine whatever, that may be recommended to them by the self-constituted doctors and doctoresses who are so frequently met with, and who pretend to possess infallible remedies for the cure of every disease. However simple some of their prescriptions may appear to be, it often happens that they are productive of much mischief, and in all cases they are injurious, by contravening the plan of treatment adopted by the physician.
7. A patient should, if possible, avoid even the *friendly visits of a physician* who is not attending him – and when he does receive them, he should never converse on the subject of his disease, as an observation may be made, without any intention of interference, which may destroy his confidence in the course he is pursuing, and induce him to neglect the directions prescribed to him. A patient should never send for a consulting physician without the express consent of his own medical attendant. It is of great importance that physicians should act in concert; for, although their modes of treatment may be attended with equal success when employed singly, yet conjointly they are very likely to be productive of disastrous results.
8. When a patient wishes to dismiss his physician, justice and common courtesy require that he should declare his reasons for so doing.
9. Patients should always, when practicable, send for their physician in the morning, before his usual hour of going out; for, by being early aware of the

visits he has to pay during the day, the physician is able to apportion his time in such a manner as to prevent an interference of engagements. Patients should also avoid calling on their medical adviser unnecessarily during the hours devoted to meals or sleep. They should always be in readiness to receive the visits of their physician, as the detention of a few minutes is often of serious inconvenience to him.

10. A patient should, after his recovery, entertain a just and enduring sense of the value of the services rendered him by his physician; for these are of such a character, that no mere Pecuniary acknowledgment can repay or cancel them.

CHAPTER II. OF THE DUTIES OF PHYSICIANS TO EACH OTHER AND TO THE PROFESSION AT LARGE

Art. I – *Duties for the Support of Professional Character*

1. Every individual, on entering the profession, as he becomes thereby entitled to all its privileges and immunities, incurs an obligation to exert his best abilities to maintain its dignity and honor, to exalt its standing, and to extend the bounds of its usefulness. He should therefore observe strictly, such laws as are instituted for, the government of its members; should avoid all contumelious and sarcastic remarks relative to the faculty, as a body; and while, by unwearied diligence, he resorts to every honorable means of enriching the science, he should entertain a due respect for his seniors, who have, by their labors, brought it to the elevated condition in which he finds it.
2. There is no profession, from the members of which greater purity of character and a higher standard of moral excellence are required, than the medical; and to attain such eminence, is a duty every physician owes alike to his profession, and to his patients. It is due to the latter, as without it he cannot command their respect and confidence; and to both, because no scientific attainments can compensate for the want of correct moral principles. It is also incumbent upon the faculty to be temperate in all things, for the practice of physic requires the unremitting exercise of a clear and vigorous understanding; and, on emergencies for which no professional man should be unprepared, a steady hand, an acute eye, and an unclouded head, may be essential to the well-being, and even life, of a fellow creature.

3. It is derogatory to the dignity of the profession, to resort to public advertisements or private cards or handbills, inviting the attention of individuals affected with particular diseases – publicly offering advice and medicine to the poor gratis, or promising radical cures; or to publish cases and operations in the daily prints, or suffer such publications to be made; – to invite laymen to be present at operations – to boast of cures and remedies – to adduce certificates of skill and success, or to perform any other similar acts. These are the ordinary practices of empirics, and are highly reprehensible in a regular physician.
4. Equally derogatory to professional character is it, for a physician to hold a patent for any surgical instrument, or medicine; or to dispense a secret nostrum, whether it be the composition or exclusive property of himself or of others. For, if such nostrum be of real efficacy, any concealment regarding it is inconsistent with beneficence and professional liberality; and, if mystery alone give it value and importance, such craft implies either disgraceful ignorance, or fraudulent avarice. It is also reprehensible for physicians to give certificates attesting the efficacy of patent or secret medicines, or in any way to promote the use of them.

Art. II – *Professional Services of Physicians to Each Other*

1. All practitioners of medicine, their wives, and their children while under the paternal care, are entitled to the gratuitous services of any one or more of the faculty residing near them, whose assistance may be desired. A physician afflicted with disease is usually an incompetent judge of his own case; and the natural anxiety and solicitude which he experiences at the sickness of a wife, a child, or any one who by the ties of consanguinity is rendered peculiarly dear to him, tend to obscure his judgment, and produce timidity and irresolution in his practice. Under such circumstances, medical men are peculiarly dependent upon each other, and kind offices and professional aid should always be cheerfully and gratuitously afforded. Visits ought not, however, to be obtruded officiously; as such unasked civility may give rise to embarrassment, or interfere with that choice on which confidence depends. But, if a distant member of the faculty, whose circumstances are affluent, request attendance, and an honorarium be offered, it should not be declined; for no pecuniary obligation ought to be imposed, which the party receiving it would wish not to incur.

Art. III – *Of the Duties of Physicians as Respects Vicarious Offices*

1. The affairs of life, the pursuit of health, and the various accidents and contingencies to which a medical man is peculiarly exposed, sometimes require him temporarily to withdraw from his duties to his patients, and to request some of his professional brethren to officiate for him. Compliance with this request is an act of courtesy, which should always be performed with the utmost consideration for the interest and character of the family physician, and when exercised for a short period, all the pecuniary obligations for such service should be awarded to him. But if a member of the profession neglect his business in quest of pleasure and amusement, he cannot be considered as entitled to the advantages of the frequent and long-continued exercise of this fraternal courtesy, without awarding to the physician who officiates the fees arising from the discharge of his professional duties.

   In obstetrical and important surgical cases, which give rise to unusual fatigue, anxiety and responsibility, it is just that the fees accruing therefrom should be awarded to the physician who officiates.

Art. IV – *Of the Duties of Physicians in Regard to Consultations*

1. A regular medical education furnishes the only presumptive evidence of professional abilities and acquirements, and ought to be the only acknowledged right of an individual to the exercise and honors of his profession. Nevertheless, as in consultations, the good of the patient is the sole object in view, and this is often dependent on personal confidence, no intelligent regular practitioner, who has a license to practise from some medical board of known and acknowledged respectability, recognised by this association, and who is in good moral and professional standing in the place in which he resides, should be fastidiously excluded from fellowship, or his aid refused in consultation when it is requested by the patient. But no one can be considered as a regular practitioner, or fit associate in consultation, whose practice is based on an exclusive dogma, to the rejection of the accumulated experience of the profession, and of the aids actually furnished by anatomy, physiology, pathology, and organic chemistry.
2. In consultations, no rivalship or jealousy should be indulged; candor, probity, and all due respect, should be exercised towards the physician having charge of the case.

3. In consultations, the attending physician should be the first to propose the necessary questions to the sick; after which the consulting physician should have the opportunity to make such farther inquiries of the patient as may be necessary to satisfy him of the true character of the case. Both physicians should then retire to a private place for deliberation; and the one first in attendance should communicate the directions agreed upon to the patient or his friends, as well as any opinions which it may be thought proper to express. But no statement or discussion of it should take place before the patient or his friends, except in the presence of all the faculty attending, and by their common consent; and no *opinions* or *prognostications* should be delivered, which are not the result of previous deliberation and concurrence.
4. In consultations, the physician in attendance should deliver his opinion first; and when there are several consulting, they should deliver their opinions in the order in which they have been called in. No decision, however, should restrain the attending physician from making such variations in the mode of treatment, as any subsequent unexpected change in the character of the case may demand. But such variation and the reasons for it ought to be carefully detailed at the next meeting in consultation. The same privilege belongs also to the consulting physician if he is sent for in an emergency, when the regular attendant is out of the way, and similar explanations must be made by him, at the next consultation.
5. The utmost punctuality should be observed in the visits of physicians when they are to hold consultation together, and this is generally practicable, for society has been considerate enough to allow the plea of a professional engagement to take precedence of all others, and to be an ample reason for the relinquishment of any present occupation. But as professional engagements may sometimes interfere, and delay one of the parties, the physician who first arrives should wait for his associate a reasonable period, after which the consultation should be considered as postponed to a new appointment. If it be the attending physician who is present, he will of course see the patient and prescribe; but if it be the consulting one, he should retire, except in case of emergency, or when he has been called from a considerable distance, in which latter case he may examine the patient, and give his opinion in writing and under seal, to be delivered to his associate.
6. In consultations, theoretical discussions should be avoided, as occasioning perplexity and loss of time. For there may be much diversity of opinion concerning speculative points, with perfect agreement in those modes of practice which are founded, not on hypothesis, but on experience and observation.

7. All discussions in consultation should be held as secret and confidential. Neither by words nor manner should any of the parties to a consultation assert or insinuate, that any part of the treatment pursued did not receive his assent. The responsibility must be equally divided between the medical attendants – they must equally share the credit of success as well as the blame of failure.
8. Should an irreconcilable diversity of opinion occur when several physicians are called upon to consult together, the opinion of the majority should be considered as decisive; but if the numbers be equal on each side, then the decision should rest with the attending physician. It may, moreover, sometimes happen, that two physicians cannot agree in their views of the nature of a case, and the treatment to be pursued. This is a circumstance much to be deplored, and should always be avoided, if possible, by mutual concessions, as far as they can be justified by a conscientious regard for the dictates of judgment. But in the event of its occurrence, a third physician should, if practicable, be called to act as umpire; and if circumstances prevent the adoption of this course, it must be left to the patient to select the physician in whom he is most willing to confide. But as every physician relies upon the rectitude of his judgment, he should, when left in the minority, politely and consistently retire from any further deliberation in the consultation, or participation in the management of the case.
9. As circumstances sometimes occur to render a special consultation desirable, when the continued attendance of two physicians might be objectionable to the patient, the member of the faculty whose assistance is required in such cases, should sedulously guard against all future unsolicited attendance. As such consultations require an extraordinary portion both of time and attention, at least a double honorarium may be reasonably expected.
10. A physician who is called upon to consult, should observe the most honorable and scrupulous regard for the character and standing of the practitioner in attendance: the practice of the latter, if necessary, should be justified as far as it can be, consistently with a conscientious regard for truth, and no hint or insinuation should be thrown out, which could impair the confidence reposed in him, or affect his reputation. The consulting physician should also carefully refrain from any of those extraordinary attentions or assiduities, which are too often practised by the dishonest for the base purpose of gaining applause, or ingratiating themselves into the favor of families and individuals.

Art. V – *Duties of a Physician in Cases of Interference*

1. Medicine is a liberal profession, and those admitted into its ranks should found their expectations of practice upon the extent of their qualifications, not on intrigue or artifice.
2. A physician in his intercourse with a patient under the care of another practitioner, should observe the strictest caution and reserve. No meddling inquiries should be made; no disingenuous hints given relative to the nature and treatment of his disorder; nor any course of conduct pursued that may directly or indirectly tend to diminish the trust reposed in the physician employed.
3. The same circumspection and reserve should be observed, when, from motives of business or friendship, a physician is prompted to visit an individual who is under the direction of another practitioner. Indeed, such visits should be avoided, except under peculiar circumstances; and when they are made, no particular inquiries should be instituted relative to the nature of the disease, or the remedies employed, but the topics of conversation should be as foreign to the case as circumstances will admit.
4. A physician ought not to take charge of or prescribe for a patient who has recently been under the care of another member of the faculty in the same illness, except in cases of sudden emergency, or in consultation with the physician previously in attendance, or when the latter has relinquished the case or been regularly notified that his services are no longer desired. Under such circumstances, no unjust and illiberal insinuations should be thrown out in relation to the conduct or practice previously pursued, which should be justified as far as candor, and regard for truth and probity will permit; for it often happens, that patients become dissatisfied when they do not experience immediate relief, and, as many diseases are naturally protracted, the want of success, in the first stage of treatment, affords no evidence of a lack of professional knowledge and skill.
5. When a physician is called to an urgent case, because the family attendant is not at hand, he ought, unless his assistance in consultation be desired, to resign the care of the patient to the latter, immediately on his arrival.
6. It often happens, in cases of sudden illness, or of recent accidents and injuries, owing to the alarm of friends, that a number of physicians are simultaneously sent for. Under these circumstances, courtesy should assign the patient to the first who arrives, who should select from those present, any additional assistance that he may deem necessary. In all such cases, however, the practi-

tioner who officiates should request the family physician, if there be one, to be called, and, unless his further attendance be requested, should resign the case to the latter on his arrival.

7. When a physician is called to the patient of another practitioner, in consequence of the sickness or absence of the latter, he ought, on the return or recovery of the regular attendant, and with the consent of the patient, to surrender the case.
8. A physician, when visiting a sick person in the country, may be desired to see a neighboring patient who is under the regular direction of another physician, in consequence of some sudden change or aggravation of symptoms. The conduct to be pursued on such an occasion is to give advice adapted to present circumstances; to interfere no farther than is absolutely necessary with the general plan of treatment; to assume no future direction, unless it be expressly desired; and, in this last case, to request an immediate consultation with the practitioner previously employed.
9. A wealthy physician should not give advice *gratis* to the affluent; because his doing so is an injury to his professional brethren. The office of a physician can never be supported as an exclusively beneficent one; and it is defrauding, in some degree, the common funds for its support, when fees are dispensed with, which might justly be claimed.
10. When a physician who has been engaged to attend a case of midwifery is absent, and another is sent for, if delivery is accomplished during the attendance of the latter, he is entitled to the fee, but should resign the patient to the practitioner first engaged.

Art. VI – *Of Differences between Physicians*

1. Diversity of opinion, and opposition of interest, may, in the medical, as in other professions, sometimes occasion controversy and even contention. Whenever such cases unfortunately occur, and cannot be immediately terminated, they should be referred to the arbitration of a sufficient number of physicians, or a *court-medical*.

   As peculiar reserve must be maintained by physicians towards the public, in regard to professional matters, and as there exist numerous points in medical ethics and etiquette through which the feelings of medical men may be painfully assailed in their intercourse with each other, and which cannot be understood or appreciated by general society, neither the subject-matter of such differences nor the adjudication of the arbitrators should be made public, as publicity in a case of this nature may be personally injurious to the individuals concerned, and can hardly fail to bring discredit on the faculty.

Art. VII – *Of Pecuniary Acknowledgments*

1. Some general rules should be adopted by the faculty, in every town or district, relative to *pecuniary acknowledgments* from their patients; and it should be deemed a point of honour to adhere to these rules with as much uniformity as varying circumstances will admit.

CHAPTER III. OF THE DUTIES OF THE PROFESSION TO THE PUBLIC, AND OF THE OBLIGATIONS OF THE PUBLIC TO THE PROFESSION

Art. I – *Duties of the Profession to the Public*

1. As good citizens, it is the duty of physicians to be ever vigilant for the welfare of the community, and to bear their part in sustaining its institutions and burdens: they should also be ever ready to give counsel to the public in relation to matters especially appertaining to their profession, as on subjects of medical police, public hygiene, and legal medicine. It is their province to enlighten the public in regard to quarantine regulations, the location, arrangement, and dietaries of hospitals, asylums, schools, prisons, and similar institutions, in relation to the medical police of towns, as drainage, ventilation, &c. – and in regard to measures for the prevention of epidemic and contagious diseases; and when pestilence prevails, it is their duty to face the danger, and to continue their labors for the alleviation of the suffering, even at the jeopardy of their own lives.
2. Medical men should also be always ready, when called on by the legally constituted authorities, to enlighten coroners' inquests and courts of justice, on subjects strictly medical, – such as involve questions relating to sanity, legitimacy. murder by poisons or other violent means, and in regard to the various other subjects embraced in the science of Medical Jurisprudence. But in these cases, and especially where they are required to make a post-mortem examination, it is just, in consequence of the time, labor and skill required, and the responsibility and risk they incur, that the public should award them a proper honorarium.
3. There is no profession, by the members of which, eleemosynary services are more liberally dispensed, than the medical; but justice requires that some

limits should be placed to the performance of such good offices. Poverty, professional brotherhood, and certain public duties referred to in section 1 of this chapter, should always be recognised as presenting valid claims for gratuitous services; but neither institutions endowed by the public or by rich individuals, societies for mutual benefit, for the insurance of lives or for analogous purposes, nor any profession or occupation, can be admitted to possess such privilege. Nor can it be justly expected of physicians to furnish certificates of inability to serve on juries, to perform militia duty, or to testify to the state of health of persons wishing to insure their lives, obtain pensions, or the like, without a pecuniary acknowledgment. But to individuals in indigent circumstances, such professional services should always be cheerfully and freely accorded.

4. It is the duty of physicians, who are frequent witnesses of the enormities committed by quackery, and the injury to health and even destruction of life caused by the use of quack medicines, to enlighten the public on these subjects, to expose the injuries sustained by the unwary from the devices and pretensions of artful empirics and impostors. Physicians ought to use all the influence which they may possess, as professors in Colleges of Pharmacy, and by exercising their option in regard to the shops to which their prescriptions shall be sent, to discourage druggists and apothecaries from vending quack or secret medicines, or from being in any way engaged in their manufacture and sale.

Art. II – *Obligations of the Public to Physicians*

1. The benefits accruing to the public directly and indirectly from the active and unwearied beneficence of the profession, are so numerous and important, that physicians are justly entitled to the utmost consideration and respect from the community. The public ought likewise to entertain a just appreciation of medical qualifications; – to make a proper discrimination between true science and the assumption of ignorance and empiricism, to afford every encouragement and facility for the acquisition of medical education – and no longer to allow the statute books to exhibit the anomaly of exacting knowledge from physicians, under liability to heavy penalties, and of making them obnoxious to punishment for resorting to the only means of obtaining it.

CHAPTER 3

STANLEY JOEL REISER

CREATING A MEDICAL PROFESSION IN THE UNITED STATES: THE FIRST CODE OF ETHICS OF THE AMERICAN MEDICAL ASSOCIATION

In 1846, a group of about 100 American doctors convened in New York City a national convention to change medical education, to produce a body of standards that would demarcate the ideals of the "regular" medical profession from that of sectarians (such as homeopaths) who threatened its hegemony, and to restore a luster that abuses of education and practice had tarnished. These reforms were proposed as the most significant policies of the organization being founded then to carry them out – the American Medical Association (AMA).

A *Code of Ethics* emerged a year later and was a great success. Between 1847, when the code ([1]) was written, and 1855, when the AMA decreed all state and local societies had to follow the code, it had become widely known and adopted throughout the United States [3]. Rightly, the preface to the code credits much of its substance to the work of Thomas Percival. But as Robert Baker shows, Percival's class-conscious language was transformed by the egalitarian culture of nineteenth century America. This culture promoted equal treatment of patients, and replaced Percival's views of a doctor-patient relationship influenced by government with one that existed in a world of personal privacy. Percival is Americanized in the AMA Code [2]. A careful examination of the *Code's* content thus reveals how mid-nineteenth century American doctors viewed their medical life and responsibilities, and what ethical principles they selected to guide their mutual relations and identity as a professional group.

The *Code* divides its comments into three main sections: the obligations of doctor and patient toward each other, the relationship among physicians themselves, and the interaction of the public and the profession. What will strike the modern mind as a great point of difference between present-day codes and

R. Baker (ed.), The Codification of Medical Morality, 89–103.
© 1995 *Kluwer Academic Publishers. Printed in the Netherlands.*

the 1847 code is the burden the 1847 code placed on patients relationships to physicians. More space is given to these obligations than to those of physician to patient. But it is with the doctor's duties to those under their care that the code begins, and when analyzing codes, significance accrues to what is stated first. "Firstness" generally implies consequence in historic documents.

THE BEGINNING

The first sentence of this *Code* is: "A Physician should not only be ever ready to obey the calls of the sick, but his mind ought also to be imbued with the greatness of his mission, and the responsibility he habitually incurs in its discharge." The passage strikes a refrain for what will follow. There is a grave sense of duty placed on the physician to respond to patient need, and to recognize that a recurrent obligation to meet it will be a constant of medical life. But the sentence is also self-regarding: the capital "P" physician throughout his work with the sick is to feel stirred by its significance of his work – the sense of mission formed in terms of conceit clouds the sense of service this opening sentence evokes. This pairing of egotism and altruism, of self-approval and self-denial appears throughout the code. It reflects a view that ethical obligations to others and expectations for personal gain are to live side by side in the professional house constructed by the AMA *Code*.

CONFIDENTIALITY

Confidentiality is specified as an important ethical concern, not only in relation to medical facts about a patient but in all personal circumstances learned about through the therapeutic relationship. The injunction to keep such disclosures secret is made emphatic by the observation that confidentiality has been recognized by courts as a significant and appropriate action by physicians.

THE VISIT

Codes commonly discuss not only principles worthy of safeguard, but also behaviors important to condemn – such as the unnecessary visit. These are criticized as provoking the anxiety of the patient and diminishing the doctor's authority by creating mistrust in the relationship. But as the code states, when an illness requires monitoring, it is important to see the patient often, even though such visits may be difficult for the doctor – as when the distance to be traveled is great. Since home care was the main locus of nineteenth-century medical treatment, the etiquette and ethics of "the visit" were important.

DISCLOSURE

One of the longest-lived ethical debates in medicine concerns the content of the disclosure about illness given to a patient by a doctor. In the Hippocratic treatise *Decorum*, the writer instructs physicians to perform their duties "calmly and adroitly, concealing most things for the patient while you are attending to him." Therapy was to be given with a serene and cheerful face, and nothing revealed to the patient of his present condition or possible future status for, Hippocratic experience had shown, "many patients through this cause have taken a turn for the worse" [5].

Over 2000 years later, the first AMA code citing (with some addition) the exact words of Percival, continues the Hippocratic perspective on this subject. "Gloomy prognostications" were condemned, and patients, if at all possible, were to be kept in the dark about threatening possibilities. The doctor was to be "the minister of hope and comfort to the sick." Disclosure was to be given to friends of the patient; if it became necessary for the patient to learn the facts, doctors were urged to withdraw in favor of laymen, from whose lips a grave prognosis would not appear so threatening.

This tradition recognized the gravity of words in medicine: they could wound as deeply as physical acts. A depressed spirit was to be avoided as much as a depressed pulse. Hope was viewed by the Hippocratics and their nineteenth-century European successors, as a key ingredient to curative or supportive medical efforts.

ABANDONMENT

Hippocratic physicians had been accused of abandoning, or refusing to care for, desperately ill patients. Their answer to this charge was to assert that to attempt to treat patients whose disease was stronger than the therapies at hand was to denigrate the art of medicine, and the practitioners of it. As the Hippocratic essay, *The Art*, puts it: "In cases where we may have the mastery through the means afforded by a natural constitution or by an art, there we may be craftsmen, but nowhere else" [5].

The issue of abandonment continued to pursue doctors in the mid-nineteenth-century. The *Code*, however, cautions that failure to produce a potentially good outcome is not a sufficient cause for refusing to aid a patient. Departing from the Greek perspective, and adopting a Christian theme, the relief of suffering is deemed a basic therapeutic goal. In such circumstances the *Code* indicates, "attendance may continue to be highly useful to the patient and comforting to the relatives around him, even in the last period of a fatal malady, by alleviating pain and other symptoms, and by soothing mental anguish."

ADVISE IN LIFE

The duties given to physicians by the *Code* included a responsibility to detect and advise the patient "suffering under the consequences of vicious conduct." Here the doctor is urged to step out of the clinical role and help the patient pursue virtue. The role is urged because of the doctor's relation of intimacy to the life and body of the patient. This expansive conception of the physicians role is not new. Hippocratic doctors also were to keep close check on the life habits of their patients and urge changes where well-being was threatened.

The Code leaves the doctor and turns to the patient.

GIVING DOCTORS THEIR DUE

Fewer obligations are enjoined on physicians as they relate to patients, than are enumerated for patients to keep in their exchange with doctors. These duties of patients begin with a requirement that they retain what is called a "just sense" of the hard work doctors perform on their behalf. This seems intended to put them in a proper frame of mind to fulfill the specific duties that follow.

SEEKING DOCTORS HAVING CHARACTER AND MEDICAL SKILL

The first duty mentioned is selecting as medical advisor someone who has gained a regular professional education. Even here there is room for choice. Not just any adviser will do. Preferable are doctors with "regular" habits of life, and without behaviors that would seem incompatible with professional demeanor.

At first reading this injunction seems to modern eyes somewhat frivolous and overly fastidious. But a long tradition of appropriate concern for the doctors behavior outside of the clinic exists. In the extraordinary one paragraph Hippocratic essay discussing requirements of good doctoring called *The Physician*, we are told" "The prudent man must also be careful of certain moral consideration – not only to be silent, but also of a great regularity of life, since thereby his reputation will be greatly enhanced" [5]. It was recognized that the ability of patients to trust physicians was dependent on the patient's view of the doctor's character. The best index of character was observed social behavior.

PRESERVING HEALTH BY GETTING CARE

Unified family care under the eye of a single physician is described in the *Code* as the best form in which medical care be dispensed. Acquaintance with the habits and predisposing factors that cause disease gives the physician a therapeutic edge. Further, patients are to seek advice for symptoms that may seem trivial. They are cautioned that "fatal results often supervene on the slightest accidents." True. But excessive concern for trivial symptoms also may lead to excessive care and hypochondria.

This is precisely the outcome in the short play written in the early twentieth century by Jules Romains – *Dr. Knock* [9]. In this tale, a young enterprising doctor purchases a declining practice from a retiring physician. He converts it into a thriving business by causing patients to worry that any new sensation may signal a dread illness. He gets so good at it that, as the play ends, he convinces the doctor who sold him what was a moribund practice, to enter the hospital to check on the meaning of a "suspicious" facial color.

Further, perhaps more than any other section of the Code, in this injunction to patients to seek care for minor symptoms, the physicians express a self-interest that seems greater than their patient interest.

CANDOR IN MEDICAL RELATIONSHIPS

This code is written in a period of transition concerning the fact-finding activities of medicine. Physical diagnosis was in the ascendancy, with physicians increasingly using simple technologies such as the stethoscope to extend their sense into the body of the patient. The best judgments in medicine were thought by doctors to be based on physical evidence they themselves detected, such as the sounds they heard in the chest of a patient. This evidence was replacing information based on the patient's recollections and sensations, because it was often flawed by lapses in memory, conflation of events, and the self-conscious withholding or distortion of facts [8].

Out of its concern about the reliability of the patient as historian, the *Code* declares it a moral duty for patients to strive for honesty and accuracy. There was a particular worry that from modesty female patients would not disclose to the mostly male doctors the place and possible causes of their ailments. But how to encourage forthright disclosure? The *Code* uses two arguments: one based on fear, the other confidentiality. Failure to communicate about symptoms freely might prevent timely treatment and thus cause the patient to "sink under a painful and loathsome disease." But when full disclosure was given, the patient was assured "that a medical man is under the strongest obligations of secrecy."

Once again the basis of trust in the medical relationship is argued on grounds of scientific skill and confidentiality.

USING THE DOCTOR'S TIME PROPERLY

Despite the appeal to the patient to tell all, in the next passage the *Code* indicates it really did not mean it. It is a restrained disclosure that physicians seem to want. The issue here seems to be a matter of time, the doctor's scarce resource. The framers of this *Code* appeared worried that doctors were being overburdened by the patients who spoke torrents of detail about themselves, their lives, and their illness during a medical visit. Self-control was urged on the patient, who "should never weary his physician with a tedious detail of events on matters not appertaining to his disease [or] even as it relates to his actual symptoms."

FOLLOWING DOCTORS ORDERS

This emphasis on patient's responsibilities in relating the disease history continues in the *Code*, as it turns its focus to therapy. Here the rhetoric becomes stronger. While in previous passages patients were urged to "entertain a just sense of duty," to "faithfully" communicate, and not to "weary" the doctor, here obedience is demanded. "The obedience of a patient to the prescriptions of his physician" this passage begins, "should be prompt and implicit." The need for patients to recognize medical authority finds even stronger assertion as the passage continues. The patient "should never permit his own crude opinions as to their [the prescriptions] fitness, to influence his attention to them." This paternalistic rhetoric was a response to the difficulties doctors encountered in having patients adhere to their therapeutic regimen, but it also represented the traditional view of patients as medically ignorant and thus at risk in acting on their own views of the therapy. It reflected too the physician's fear that to disclose to patients the nature of their disorder courted the danger of depressing and injuring them. The consequent barrier to communication and collegiality between doctor and patient inevitably made the relationship one of order-giver to order-taker.

KEEPING FAITH WITH YOUR DOCTOR

Patients could be endangered not only by the intrusion of their own views on the prescribed therapy, but also those of other physicians. The *Code* recognized that a variety of approaches were possible in dealing with a given illness in a given patient. Its writers were concerned that patients conversing about therapy with doctors other than their own might produce contrary views about what ought to be done. Such contradictions presented sources of alarm to patients, and might cause them to alter or stop current therapy, or lose confidence in their doctor, or (the implication was present) even change practitioners. Hence patients were urged to avoid even (and these words appear in the *Code* in italics) "*friendly visits of a physician*" not attending them, and "never" to speak of their illness if such an encounter occurs.

GRATITUDE AS A DUTY OF PATIENTS

The last obligations of patient to doctor in this section hark back to the rejoinder with which it began – to honor the obligation of gratitude for the doctor's work in the patient's behalf. Thus, when leaving a medical relationship, explain why. If possible, send for the doctor in the morning so that the day's schedule of travel is appropriately formulated. Do not interrupt the meals or sleep of doctors, if possible, by visits at these times. But always be ready to receive their visits. Pay your bill, and recognize that the doctor's services "are of such a character, that no mere pecuniary acknowledgements can repay or cancel them." This final remark gives the medical relationship the tone of a parental one.

In the second main part of the *Code* the doctor and patient relationship gives way to the mores of collegial associations among physicians, and of physicians with the medical profession itself.

A TRUSTWORTHY PUBLIC FACE

All professions have a public face. The characteristics which the *Code* seeks to engrave on the social visage with which it would meet the public are dignity, honor, temperance, and beneficence. These characteristics, the *Code*'s framers hoped, would produce two main results: "to exalt" the social standing of the medical profession and "to extend the bounds of its usefulness."

These goals, however, were threatened by a number of contingencies to which this part of the *Code* pays great attention. Members were encouraged to take quite seriously the standards of professional conduct enumerated in the *Code*. Indeed the way they put it is interesting. Although these standards and the medical association itself are agencies a given doctor voluntarily participates in, nonetheless they are characterized in the *Code* as "laws . . . instituted for the government of its members." This clearly leaves little room for disagreement or flexibility. As our subsequent commentary will show, this rigidity would be a later source of controversy.

ETHICAL BEHAVIOR AND MEDICAL EXCELLENCE

In addition to closely following the *Code*'s standards, members of the profession were to avoid public criticism of colleagues, and to treat senior physicians with a respect based on gratitude for past efforts on behalf of medicine. "Moral excellence" was urged: in this context it was observed that scientific excellence could not compensate for its absence.

The last point is exceedingly important. Its significance has been dimmed in modern times by the rise of physicians' technologic prowess. In some modern cases where physicians have trespassed upon the trust of their patients, the issue of whether they were good physicians or not mainly has been argued on technological grounds. The question was put: Were they exceedingly skilled or not in performing certain operations or diagnostic interventions? Such arguments fail to grasp a central point. If physicians cannot be trusted to act with respect for the person of their patients, they should not be permitted access to patients no matter what their technologic skill. The good physician must have *both* ethical and technologic competence: lacking either, wholeness and thus excellence is not possible.

ADVERTISEMENTS AND MEDICAL CHARACTER

"Be temperate in all things," the *Code* urges doctors so that a steady hand and clear mind would be available for emergencies. This follows consistently from the *Code*'s advice to patients, cited earlier, to give preference to doctors "whose habits of life are regular." The association of character with trust, embodied in this advice, is responsible in large measure for the strict view the *Code* takes of advertising by physicians. "It is derogatory to the dignity of the profession, to resort to public advertisements or private cards or handbills." The *Code* enumerates other forms of behavior that invite adverse public attention to doctors practices: offering free advice to the poor, promising quick cures, publishing results of cases, inviting laymen to witness operations, or promoting testimonials of patients. Advertisement was linked to the practices of the so-called "empirics" who developed and sold medicines, removed cataracts, delivered babies, and offered a host of specialized treatments for particular diseases. The discomfort felt then and today concerning advertisement concerns its purpose and associations.

When we are sick, we engage in extraordinary behavior: we visit a person who is increasingly – in this age of specialization and frequent change of residence – a stranger to us; we disrobe; we allow the stranger to prod and poke us; we reveal close held feelings and experiences, often told to no one else. We permit therapy to be given that may be disagreeable or painful, even to the point of allowing our body to be cut open, things removed, and then sewn up again. Nowhere else in our lives would such actions be sanctioned by people hardly known to us, unless we had a transcendent belief in a moral principle – that this medical stranger was committed to performing only those actions that would help us. Without such trust patients would not permit medical interventions. At the heart of medical uneasiness with advertisements is its connection to selling something. The person who is selling us seems more self-serving than selfless, not trustworthy enough to give ourselves over to.

THE PROPRIETY OF PATENTS

Old traditions of patenting or concealing remedies also threaten the character of the medical profession. They might deprive other doctors, or their patients, of the full benefit of significant knowledge, or, "if mystery alone give it value," they dampen the scientific credibility of medicine. Perhaps the most famous instance of this behavior occurred in the early seventeenth century when Peter Chamberlain invented the obstetric forceps. Before this technology, if the fetus proved too large to pass through the birth canal the result often was the double tragedy of maternal and fetal death. But the forceps was kept a secret within the Chamberlain family of midwives for the remainder of the century, until a member revealed the source of their success in aiding childbirth.

Thus sharing knowledge was an activity would promote the values of "beneficence and professional liberality."

DOCTORS TREATING EACH OTHER

Who is to heal the healer? When physicians get sick they are brought to recognize that having knowledge is not sufficient to produce a good outcome in medicine. Objectivity is needed to focus their knowledge, lest its power be misdirected and harm produced. The *Code* recognizes the difficulty physicians experience when either self or loved one fall sick. To shield them from the dangers of this situation, it asks practitioners to serve each other gratuitously. Doctors are also asked to substitute for one another when circumstances, such as illness, prevent them from treating their patients. These situations are crucial to drawing the attention of doctors to another facet of their interdependence – that of shielding each other from the painful dilemma of having knowledge but being unable to responsibly act.

THE ETHICS OF CONSULTATION

One of the great events in medicine occurs when physicians reveal to a patient that they do not know enough to deal with a problem and need the help of a colleague who does. It is the time when the authoritative doctor publicly admits ignorance and patients have at the disposal the wisdom of the profession at large to solve the puzzle of their illness. Consultation is one of the most significant actions in medicine, and also one of the most difficult. Many obstacles exist along the avenue between the referring and the consulting physician, any of which might become a significant barrier. Accordingly, more space is devoted in the *Code* to discussion of consultation than to any other single issue.

The *Code* begins its main work on this subject by reminding physicians that only those with a regular medical education are fitted to the role of consultant. Having disposed of that issue, it develops an elaborate set of guidelines to deal with the relationships of consultation. In the meeting together of attending (the doctor in charge of the case) and consulting physicians, the attending should question the patient first, be the one to give the results of the consultation to the patient, and be free to vary the treatment recommended by consultants.[1] When jointly visiting patients, should the consultant arrive first, he must wait for the attending, and postpone the examination if the attending fails to arrive.

The *Code* pays a great deal of attention to the problem of disagreement between the medical parties. The view is taken that controversy on the way to reaching a judgment should remain concealed from the patient and that responsibility for the success or failure of the therapy chosen should be shared equally. However, should disagreements not be reconciled, another physician is to be called in as umpire. After this, should any remaining doctor be unwilling to agree with the rest, withdrawal from participation is warranted. There is great concern, expressed in several places in the *Code*, that disagreements made public bring discredit to the profession, for the public was likely to misunderstand the terms of the dispute.

The final concern was how the consultant spoke of the attending in front of patient and family. Care was to be taken not to wound the attending's reputation, nor to be overly solicitous. The interest here was that the consultant not take advantage of the entry to patients provided by the attending and steal them. This is a crucial admonition. The avenue of consultation would be closed were such a problem to arise.

MEDICAL COMPETITION

The last discussion reveals a possessiveness of patient, which has both a selfless and self-regarding tone. Doctors are warned against this behavior when they meet the patients of others in daily life, or visit the home of a sick friend or acquaintance. If an emergency brings several physicians to a home, the first to arrive assumes authority. When treating the patient of an absent colleague, on recovery, the patient is returned (with consent) to the original doctor.

THE PUBLIC AND THE PROFESSION

The third basic division of the *Code* discusses the relation between the medical profession and the public. The services physicians ought to give their community as a consequence of their medical knowledge are listed. These include advice about: quarantine, public hygiene, medico-legal issues, the location and content of hospitals, appropriate drainage and ventilation in institutions and the town, biologic evidence in trials, and keeping the public from taking "quack medicines," even to the point of boycotting druggists who sold such remedies. Public duty required physicians to risk their lives when epidemics threatened the community, and to proffer their services to those in need.

However the *Code* places boundaries on charitable obligations. There is concern, for example, that richly endowed private or public charities not take advantage of the doctor's good will and request *pro bono* care. This was provided only to impoverished individuals to whom "professional services should always be cheerfully and freely accorded." When fees were charged, the *Code* recommended that they be based on schedules drawn up by different localities.

REACTIONS TO THE CODE

Arrayed against the AMA *Code*'s position that group standards and governance are essential tools in shaping practice to fit the needs of patients and society is the position staked out a century earlier by John Gregory in his 1772 *Lectures on the Duties and Qualifications of a Physician* [4]. Gregory urged not a professional union, but the education of a literate segment of learned laymen in medicine as the best means of preserving good practice. Such medically enlightened laymen could help the public recognize superior practitioners and encourage their success; they could also help spread medical knowledge useful in emergencies and other facets of medical care. Leaving patients ignorant and dependent on the medical expert left them hostage to professional courtesies, which made distinctions among physicians based on merit difficult to air publicly. Further, the profession's interest in securing medical attendance only by practitioners of the "regular" medicine meant for Gregory, that potentially helpful unorthodox remedies or practitioners would not be used. He wrote:

> It is a physician's duty to do everything in his power that is not criminal to save the life of his patient, and to search for remedies from every source and from every hand, however mean and contemptible. This, it may be said, is sacrificing the dignity and interests of the faculty. But I am not here speaking of the private policies of a corporation, or of the little arts of a craft. I am treating of the duties of a liberal profession whose object is the life and health of the human species. . . . the dignity of which can never be supported by means that are inconsistent with its ultimate object [4].

A second viewpoint emerged several decades after the writing of the AMA Code. By the early 1880s the AMA had come to dominate medicine in the United States and its *Code* was widely accepted by local, county, and state medical societies. A group of physicians in the Medical Society of New York called into question its hegemony on ethical values and judgments applied to practice. One of them, Louis Pilcher, accused the AMA of creating a medical elite whose zeal to observe the strict letter of the *Code* caused them "to wholly ride roughshod over the rights of others when such rights are not protected by any distinct provision of the *Code*" [7]. Further, this elite was said to apply to the *Code*'s provisions unequally, with the prominent and influential practitioners escaping its discipline, and the obscure and weak expected to comply. But, in his most fundamental challenge to the *Code*, Pilcher questioned why its standards should apply to *all* members of the profession, and why doctors who refused to declare their allegiance to it should be characterized as unworthy of professional recognition. Pilcher declared: "The physician is a freeman; he has ceased to recognize paternal interference with his judgment; he wears the livery of no employer; he acknowledges the restrictions of no trades-union."

CONCLUDING REMARKS

There are four constituencies whose views are central to the proper functioning of medicine: the patient, the practitioner, the society, and the profession. Gregory put forth in the earliest focused discussion I know of the argument for the rights and warrant of patients and laymen in deciding medical issues. In contemporary times this argument for patient rights has been made forcefully and has improved decisions in medical care. However, as patients, we are limited as decision-makers by that distortion of logic imposed on us by powerful emotions and anxiety-reducing mental devices called forth during illness. Knowledge is no shield against them, as proven by the difficulties doctors have with patients. For this reason we turn to others to help structure choices and direct therapy.

The "others" we most frequently turn to, are physicians. For them one of the significant traits that leads to skillful patient care is an ability to think freely of the most appropriate approaches to treating illness in the varied beings who are their patients. Pilcher makes a strong case for physicians to choose which ethical values shall dominate in any given medical care decision. However, free choice without normative boundaries can be chaotic. The use of scientific and ethical standards to define a playing field within which actions can vary creates a good balance between anarchy and subjugation.

The role of the state in regulating medicine is thwarted by the complexity and privacy of medical actions. While complexity might be dealt with in medicine as in other complex but more regulated disciplines, such as architecture, the second factor, privacy, presents problems. So much of what takes place in medicine is in venues beyond the ken of outside observers as in the operating or consulting room. So much of medical care involves revelations that may occur only if secrecy is promised – who we really are, and the thoughts and actions of personal life – that the knowledge that detailed oversight occurs would reduce the doctor's effectiveness. Social oversight must thus have limits.

This brings us to the profession itself as an actor in medicine. Its primary role is not just to create scientific and ethical standards, but to forge an ethos that commits the practitioner to follow these standards. This was made most clear in the first major ethical code of Western medicine, the Hippocratic Oath [5]. Perhaps its most significant passage comes just after the introductory sentence, when it describes the relationship among students to each other and to their teachers. The relationship portrayed is one of the family – teachers were like fathers, and fellow students, brothers. The image of family is invoked in this passage to create bonds that will strengthen the prospect that the succeeding ethical principles enumerated in the code will be followed. This developed group conscience was to reinforce individual conscience in steering practitioners on an appropriate course. The creation of an ethical ethos among practitioners and the development of standards to form the moral boundaries of practice are critical features for the effective functioning of medicine, and an important purpose of codes and professional groups.

But critics of the medical profession and the codes that embody its standards often see their ethical assertions of protecting patients' welfare as in reality measures to further the doctors' gain. I believe this line of criticism misreads intentions and ignores outcomes. The profession of medicine has developed to serve both patient and practitioner. The critical issue is whether benefit to practitioner occurs only from the circumstance of meeting the true needs of and serving the patient. The purpose of the first AMA code is predominantly in this direction.

In modern times, we recognize that we are better off if the informed patient sought for by Gregory, the more independent doctor called for by Pilcher, and a more activist society interested in the goals and methods of medicine join with the medical professions in a balanced manner to decide the course of health care. But as a beginning toward this more complicated arrangement in the United States, the first code of the American Medical Association was a source of benefit and a beacon of appropriate change.

NOTE

[1] The use of the term 'attending physician' to refer to the doctor in charge of the case, as distinguished from the *consultant*, is made explicitly in the 1847 AMA *Code*. While it may not be the first explicit use of this term, it is an early formulation of the concept. In Percival's 1803 *Medical Ethics* [6], the principal doctor is referred to as the "physician in attendance." We see in the AMA *Code*, which was heavily based on Percival, the transmutation from the phrase "in attendance" to "attending."

BIBLIOGRAPHY

1. American Medical Association: 1847, "Code of Ethics," *Minutes of the Proceedings of the National Medical Convention held in the City of Philadelphia, in May 1847*, pp. 83–106; this volume, pp. 75–88.
2. Baker, R.: 1993, "Deciphering Percival's Code," in Baker, R., Porter, D., and Porter, R (eds.), *The Codification of Medical Morality: Historical and Philosophical Studies of the Formalization of Western Medical Morality in the Eighteenth and Nineteenth Centuries: Volume One: Medical Ethics and Etiquette in the Eighteenth Century*, pp. 179–212, Kluwer Academic Publishers, Dordrecht.
3. Burns, C.: 1977, "Reciprocity in the Development of Anglo-American Medical Ethics, 1765–1865," in Burns, C. (ed.), *Legacies in Ethics and Medicine*, pp. 300–6, Science History Publications, New York; this volume, pp. 135–146.
4. Gregory, J.: 1772, *Observation on the Duties and Offices of a Physician and on the Method of Prosecuting Enquries in Philosophy*, W. Straham and T. Cadell, London.
5. Jones, W. H. S.: 1923, *Hippocrates*, Harvard University Press, Cambridge, MA, Vol. I (trans.).
6. Percival, T.: 1803, *Medical Ethics; Or, A Code of Institutes and Precepts, Adapted to the Professional Conduct of Physicians and Surgeons*, J. Johnson, London.
7. Pilcher, L.: 1883, *Codes of Medical Ethics in An Ethical Symposium: Being a Series of Papers Concerning Medical Ethics and Etiquette from a Liberal Standpoint*, G P Putnam, New York.
8. Reiser, S.: 1978, *Medicine and the Reign of Technology*, Cambridge University Press, Cambridge.
9. Romains, J.: 1925, *Dr. Knock: A Comedy in Three Acts*, S. French, London.

CHAPTER 4

TOM L. BEAUCHAMP

WORTHINGTON HOOKER ON ETHICS IN CLINICAL MEDICINE

Throughout ancient, medieval, and modern medicine, the physician's moral obligations and virtues have been conceived primarily through professional commitments to provide care, expressed as fundamental obligations of beneficence. The physician must maximize the patient's medical benefits above all competing obligations. Practices of truthtelling, confidentiality, and all aspects of patient care are, from this perspective, governed by a *beneficence model* of primary responsibility.

In the last two decades the rival idea has emerged that the proper model of the physician's primary responsibility is not medical beneficence but the autonomy rights of patients, including their rights to truthfulness, confidentiality, privacy, disclosure, and consent. This challenge has jolted medicine from its traditional preoccupation with a medical benefit in the direction of an *autonomy model* of the physician's responsibility for the patient: The physician must disclose truthfully, maintain confidentiality, and seek permission from the patient in order to respect the patient's autonomy, irrespective whether medical benefits will be increased.

Connecticut physician Worthington Hooker was the first champion in the history of medical ethics of something like an autonomy model.[1] He and Richard Clarke Cabot may have been the only physician champions of this model prior to the second half of the twentieth century. Moreover, there may never have been a figure who swam, in regard to truthtelling, so against the stream of indigenous medical tradition.

Physicians during the nineteenth century occasionally went on record in favor of respecting the wishes of patients and making appropriate disclosures. Such respect was typically defended on grounds that this form of communica-

R. Baker (ed.), The Codification of Medical Morality, 105–119.
© 1995 *Kluwer Academic Publishers. Printed in the Netherlands.*

tion with patients would have a beneficial therapeutic outcome. Hooker was different in asking both for more truthtelling and in justifying the request by appeal to the rights of patients and social utility, rather than by conventional appeals to medical beneficence.

Hooker's most important book on medical ethics was *Physician and Patient* [12]. It merits attention for its firm defense of truthtelling, for its departure from the first American Medical Association *Code*, and as the most original contribution to medical ethics by an American author in the nineteenth century. This book and Hooker's later works were aimed at quackery, abuses by regular physicians, and all others in society who "unjustly cast" aspersions on the medical profession to which Hooker was resolutely devoted ([12], pp. viii–ix; Chapter 4).

To understand Hooker's objectives, we need first to examine some aspects of his history, together with underlying assumptions of the ambiance in which he wrote and practiced.

THE BACKGROUND IN MEDICAL PRACTICE AND MORAL PHILOSOPHY

During a period of forty years, Hooker received an M.D. from Harvard Medical School (1829), practiced in Norwich, Connecticut (1829–1852), and served as Professor of the Theory and Practice of Medicine at Yale (1852–1867). Throughout this time, medicine in North America faced a challenge to its credibility. Hooker and many physicians were preoccupied with threats to their profession and reputation, often threats presented by quacks and sects. Even in regular medicine, few methods of treatment were standard, and open skepticism prevailed in the public's eye about medical efficacy. In response, practitioners of regular medicine – sometimes called, to Hooker's chagrin, allopathic medicine – sought to shore up its public standing and professional standards.[2]

This practical objective had, from the inception of the AMA, been intentionally packaged as part of its medical ethics. John Bell, in presenting the original *Code of Ethics* to the 1847 AMA Convention, stressed the importance of the obligation "to bear emphatic testimony against quackery in all its forms" [2]. *The Code* itself created a duty of physicians to combat "the enormities committed by quackery, and... the use of quack medicines, to enlighten the public on these subjects," and to expose "artful empirics and impostors" (Chapter III, Art. I, § 4). These statements closely resemble Hooker's declaration of "the objects for which this book was written" in the Preface to his *Physician and Patient* ([12], p. vi).[3]

Many passages in the AMA Code were transcribed verbatim from Thomas

Percival's *Medical Ethics* [21].[4] Easily the dominant influence in both British and American medical ethics of the period, Percival argued that:

To a patient . . . who makes inquiries which, if faithfully answered, might prove fatal to him, it would be a gross and unfeeling wrong to reveal the truth. His right to it is suspended, and even annihilated; because, its beneficial nature being reversed, it would be deeply injurious to himself, to his family, and to the public. And he has the strongest claim, from the trust reposed in his physician, as well as from the common principles of humanity, to be guarded against whatever would be detrimental to him. . . . The only point at issue is, whether the practitioner shall sacrifice that delicate sense of veracity, which is so ornamental to, and indeed forms a characteristic excellence of the virtuous man, to this claim of professional justice and social duty ([21], p. 166).[5]

Percival counseled physicians in bleak cases "not to make gloomy prognostications... but to give to the friends of the patients timely notice of danger . . . and even to the patient himself, if absolutely necessary" (Chapter II, Article 3, [21], p. 31; [18], p. 91). On the other hand, he warned specifically, as had Benjamin Rush before him, that to silence a patient with blunt authority may only result in a worsening of the patient's condition in less grave situations ([18], p. 91; [21], p. 31; see also [19]). These passages appeared almost verbatim in the AMA *Code*, as its statement of the obligations of physicians in regard to truthtelling (Chapter I, Art. I, § 4).

Percival was struggling in these passages against the arguments of his friend, the Rev. Thomas Gisborne, who opposed practices of giving false assertions intended to raise patients' hopes and lying for the patient's benefit: "The physician . . . is invariably bound never to represent the uncertainty or danger as less than he actually believes it to be"[6] ([9], p. 401). From Percival's perspective, the physician does not lie in beneficent acts of deception and falsehood, as long as the objective is to give hope to the dejected or sick patient. The role of the physician, he said (and the AMA repeated in its *Code*), is primarily to "be the minister of hope and comfort" ([21], Chapter II, Article 3, [21], pp. 31, 156–68; [18], pp. 91, 186–90). Percival was concerned about the appearance and the consequences of acts of deception, because they sometimes endangered the gentlemanly image of the physician, the potential recovery of the patient, and the character of the physician as a moral agent. But, overall, he thought the benefits outweighed the harms of such deception.

While examining Gisborne's stern admonitions on truthfulness, Percival consulted Francis Hutcheson, then considered a leading authority in moral philosophy. Percival was pleased to find Hutcheson teaching that benevolent deception in medicine is often the manifestation of a virtue, rather than an act constituting an injury:

No man censures a physician for deceiving a patient too much dejected, by expressing good hopes of him; or by denying that he gives him a proper medicine which he is foolishly prejudiced against: the patient afterwards will not reproach him for it. – Wise men allow this liberty to the physician in whose skill and fidelity they trust: Or, if they do not, there may be a just plea from necessity ([21], pp. 162–164].[7]

Hutcheson's eighteenth-century paternalism was equaled by that of the most probing British moral philosopher of the nineteenth century, Henry Sidgwick, who held that veracity can be justifiably overridden by beneficence: "Where deception is designed to benefit the person deceived, Common Sense seems to not hesitate to concede that it may sometimes be right: for example, most persons would not hesitate to speak falsely to an invalid, if this seemed the only way of concealing facts that might produce dangerous shock; nor did I perceive that any one shrinks from telling [certain] fictions to children" ([22], pp. 315–316).

Hooker's general ethical theory was in many respects similar to that of his contemporary Sidgwick, but Hooker would have been disappointed by Sidgwick's justification of benevolent deception, had he lived another seven years to witness the publication of Sidgwick's *Methods of Ethics* (1874). A few years before Sidgwick published these views, Hooker had devoted a section of his work on truthfulness to rebutting the precise point advanced by Sidgwick (as it applied to paternalism with children).

HOOKER'S RELATIONSHIP TO PERCIVAL AND TO THE AMA CODE

The American Medical Association (AMA) accepted virtually without modification the Hutcheson-Percival paradigm in its 1847 *Code*. The *Code*, and most codes of medical ethics before and since, do not include rules of veracity. The Hippocratic writings did not impose obligations of veracity, nor did the influential twentieth-century *Declaration of Geneva of the World Medical Association*, nor did any of the *Principles of Medical Ethics* of the American Medical Association in effect from 1847 to 1980. The typical maneuver has been to say that member physicians have discretion over and should exercise judgment about what to divulge to patients.

Hooker rejected all such compromises on truthtelling, at the same time putting himself forward as among the committed adherents of the AMA *Code*. He reprinted the Code in *Physician and Patient* and warmly endorsed it as a practical instrument for uniting educated physicians and for promoting medical education. He gave two reasons for his support. The primary justification was instrumental and regulative. Hooker saw the *Code* as "a great and a permanent agency in the overthrow of empiricism" ([12], pp. 256–7), that is, quackery. A connected but secondary reason was the Code's potential for "the elevation and advancement of our common calling" ([12], pp. 256–7).

Although he defended the AMA's general objectives in issuing the *Code*, Hooker refused to budge on truthtelling. Hooker never denied the value of a beneficence model, kept within proper limits. He even attempted to delineate the nature and proper boundaries of the physician's principled commitment to do good for and prevent harm to patients (see [16], p. 43). However, these goals of therapeutics were, he thought, misplaced when transplanted to the medical ethics of disclosure.

Hooker's attack on the paternalistic currents of his times was directly mounted against Percival, who Hooker took to represent "the views of those who are in favor of an occasional departure from truth" ([12], pp. 357–60). The passages in Percival suggesting the justifiability of benevolent deception of patients and the absence of a right in the patient to the truth were entirely unsatisfactory to Hooker. The medical cases Percival put forward to illustrate or support his point were judged by Hooker to be "of the most egregious character, and yet they are fair representations of the kind of deception which many feel authorized to use in the sick room" ([12], pp. 379–80).

ARGUMENTS AGAINST BENEVOLENT DECEPTION

Hooker's arguments for the obligation of veracity are primarily consequentialist, or, as Hooker puts it, they are arguments from the principle of expediency ("expediency" here meaning the most suitable or appropriate means to a justifiable end, [12], p. 360).

Perhaps the main Percivalean argument that Hooker set out to refute prescribed nondisclosure of ruinous diagnoses and prognoses in the therapeutic setting on grounds that the patient would be caused counterproductive anxiety or direct harm, thereby violating the physician's obligations of beneficence and nonmaleficence. Hooker countered by arguing that the underlying claims of hurtfulness from disclosures are not warranted by clinical experience, when the physician has consistently pursued a course of frank and candid discussion. Hooker argued that a nondeceptive means of discussion is generally more satisfactory than a deceptive means. Even when negative reactions to bad news do occur, the effects are not usually as serious to the patient, in Hooker's judgment, as the patient's reaction upon discovery or suspicion of deception by physicians ([12], pp. 361–5).

Another assumption of benevolent deception is that the concealment of truth can be effectively carried out by the physician. This belief is both precarious and dangerous, in Hooker's judgment. Even careful plans to conceal the truth are often recognized as such by patients, and many physicians do not have the acting skill to effectively deceive patients for long. Once the deception is detected, the deceived patient feels cheated, and serious injuries may, as a result, be caused to the person. These injuries may also irreparably damage the patient-physician relationship ([12], pp. 361–6).

Hooker keeps his argument close to the framework of consequentialism. A striking portion centers on the problem of risk to trust and confidence in the physician-patient relationship. He views trust and confidence as essential to any meaningful relationship of cooperation and to any positive influence that might occur from the physician's advice and control. Deception risks this loss and, if repeatedly practiced, may altogether undermine the patient's trust and willingness to act on (even correct) information provided by physicians ([12], pp. 366–72).

Hooker's most impressive argument builds on the above arguments through an appeal to "the *general effect* of deception." This argument is not concerned with the individuals subjected to deception, but rather with the impact of deception on "the whole mass of society." Here Hooker uses an argument that we might today identify as rule-utilitarian: Although deception will in some (inherently unpredictable) cases produce a momentary or an overall good for the patient, the general and remote results across the whole setting of medical practice and patient care will eventuate in "vast and permanent evils" that will never be counterbalanced by any positive results ([12], p. 372). Deception is a poison in the stomach of medicine.

In effect, Hooker invokes his principle of expediency to argue that the risks of deception outweigh the possible benefits.

> The good, which may be done by deception in a *few* cases, is almost as nothing, compared with the evil which it does in *many* cases. . . . And when we add to this the evil which would result from a *general* adoption of a system of deception, the importance of a strict adherence to truth in our intercourse with the sick, even on the ground of expediency, becomes incalculably great ([12], pp. 378–9).

Hooker further points out that if deception were to become a settled policy or prevailing practice among physicians, as "an acknowledged common rule," general distrust would inevitably ensue in the patient population, where the rule could not be kept secret. The rule of deception would thereby work to defeat the very purpose of the rule, because deception could no longer be deceptive. No one could be deceived because they would never believe they had been told the truth ([12], pp. 375–6).

The principle of expediency is here applied not to particular *actions* of deception but rather to the justification of *rules* of conduct that determine whether acts are right or wrong. Actions can be justified only by appeal to rules such as "Don't deceive." These rules, in turn, are justified by appeal to the principle of expediency, that is, utility. Hutcheson and Percival, by contrast, justified benevolent deception by direct appeal to the consequences, without the buffer of rules.

Hooker's opponents might reply that if a patient's health may on some occasions maximally be advanced through deception, then a rule of truthtelling should be only occasionally disobeyed, so that we will be better off in the moral life if we sometimes obey and sometimes disobey publicly-advocated rules. In effect, the claim is that the concealment of truth is now and then permissible, because it will not fundamentally erode either moral rules, our general respect for morality, or trust in the physician.

Hooker replied (to roughly this argument) that the use of occasional deception in a few, well selected "urgent cases" will not work. Sympathetic though he was to a utilitarian perspective,[8] Hooker denied that physicians can simultaneously predict (with success) the most beneficial outcomes in particular cases, carry through the deception without discovery by patients, and practice such deception without its being generally known that it has become the practice in the medical community. Once a patient knows that deception is permitted in the system, the patient will commonly suspect that his or her case falls under this exceptive rule.[9] In effect, Hooker met the act-utilitarian argument found in Percival with a rule-utilitarian construction.

From this series of arguments he concluded that deception should not be practiced at all in medicine. No other conclusion, he thought, was consistent with the principle of truth.

THE WITHHOLDING PROVISO

Despite these stern defenses of truthtelling, Hooker allowed some suppression of information under conditions similar to those sanctioned by Percival. If the physician is uncertain about the matter, or, if the information to be disclosed is likely to confuse the patient without the possibility of clarification, then, Hooker reasoned, disclosure would itself amount to deception, and the physician's fundamental objective must be to prevent deception ([12], p. 381).

Hooker was keen to be understood as confining his theses and arguments to whether "real falsehood is justifiable" and not (1) whether benevolent deception is ill-motivated or (2) whether the truth can sometimes justifiably be withheld. He acknowledges both that benevolent deception might spring from the best and kindest motives and that the truth can in some cases be justifiably withheld ([12], p. 360). The latter is a proviso on his otherwise firm rejection of a lack of truthfulness, and this withholding proviso deserves further examination.

Like Sidgwick, Hooker regarded some forms of disclosure as nonobligatory and subject to discretion. He attempted to distinguish between veracity or truthfulness, on the one hand, and withholding information on the other hand. Perfect conformity to all known or believed fact is only a prima facie obligation, in his reckoning. If there is good reason for withholding, this may be done, although deception and lying cannot legitimately be placed under the province of acceptable withholding. The following conclusion is paradigmatic: "There are cases in which [withholding information] should be done. All that I claim is this – that in withholding the truth no deception should be practised, and that if sacrifice of the truth be the necessary price for obtaining the object, no such sacrifice should be made" ([12], p. 380).

In using the language of deception Hooker seems primarily to mean telling another person what one believes to be false in order to deceive that person. Intentionally instilling a belief in what is false is the core of what he finds abhorrent. The rules of veracity that he thought should never be violated in medicine, then, include the obligation not to deceive others, but not an obligation of full conformity to known or believed fact.

In contemporary medical ethics, it is often held that intentional deception that does not involve lying is generally less difficult to justify than lying, because it does not as deeply threaten the relationship of trust between deceiver and deceived as does lying. Underdisclosure and nondisclosure are still less difficult to justify, from this perspective. However, this does not seem to be Hooker's position, because of his core reliance on falsehood. If the deception involves intentionally getting the person to believe what is false, Hooker does not see the relevance of a distinction between underdisclosure, deception, and lying.

Forms of deception that stand to violate obligations of truthtelling, so understood, include giving placebos or medicines under false pretenses and intentionally deceiving by the manipulation of information. Once patients entrust their care to clinicians, patients thereby obtain a right to information that clinicians would not otherwise be obligated to provide.

Why, in light of his firm arguments against falsehood, did Hooker allow this proviso in support of withholding information? The key is found in the characteristic nineteenth-century reliance (so prominent in Percival and the

AMA's *Code*) on the category of *hope*. A skillful and judicious practice of medicine involves the delicate art of instilling hope in the patient while avoiding despondency. Within the bounds of maintaining veracity and candor, and always avoiding wide departure from the truth through false assurances, the physician may legitimately avoid an undue and sharp bluntness or abrasiveness in order to avoid extinguishing hope. Bluntly presenting the patient with the worst possible outcome is plainly not the physician's obligation and not a part of skillful practice ([12], pp. 345–52).

The facts may be trimmed, in Hooker's arguments, so long as the trimming is not deceptive: "Giving utterly false assurances to the patient is a very different thing from merely exciting the hope in his mind to such a degree as the case may allow, that the remedies will produce the desired relief. . . . [Hope's] cordial influence should always be employed, so far as it can be done consistently with truth, and no further" ([12], p. 348). The line is properly drawn, in Hooker's mind, at the "sacrifice of truth." The physician's right to withhold information stops at the point the price paid for nondisclosure is the sacrifice of truth ([12], p. 380).

Hooker's view, then, is that the patient's right to know is not sufficiently broad to include every fact that might be disclosed or even every relevant fact. The physician's obligation to disclose is contingent on whether too much disclosure causes *harm* and robs the patient of justified *hope*, by revealing an upsetting condition. Medical beneficence warrants underdisclosure so long as no deception is uttered. This form of intentional suppression of information is not, in Hooker's estimation, a violation of the patient's autonomy rights or of the fundamental duties of the physician.

INCOMPETENT PHYSICIANS IN THE GRIP OF MEDICAL DELUSIONS

Hooker followed, *Physician and Patient*, with *Lessons from the History of Medical Delusions* (1850) and *Inaugural Address as Professor of the Theory and Practice of Medicine in Yale* (1852). The *Inaugural Address* criticized physicians for deceiving patients, for providing unnecessary services as if necessary, and for studying "the science of patient-getting, to the neglect, to some extent at least, of the science of patient-curing" ([15], p. 27). The longer *Lessons* was an exploration of the delusions underlying false conjectures in medicine. Both works critiqued incompetence and delusion in medicine.

Hooker maintained that well-meaning, well-educated physicians and patients can be deluded no less than quacks. He presented a theory of error and truth in medical belief and argued against breaches of confidentiality ("medical secrecy,"

[13], pp. iv, 11, 102–5; see also [11], pp. 10–8, 20–1). Hooker labored in these works to protect both patients and physicians from the delusions that were (often honestly) presented by physicians as if they were medical truths.

Hooker observed that the judgment of physicians in making recommendations, no less than the patient's in receiving them, are subject to distortions and need to be checked by sober sources whenever there persists an underlying enthusiasm for a remedy. Among the many dispositions to error on his list was the following:

> *The disposition to adopt exclusive views and notions* . . . gives to its possessor the character of a *one idea man*. . . .

The way in which this disposition leads to error is this: A physician has his attention directed to a particular set of facts. He becomes intensely interested in them. They fill the field of his mental vision, and he becomes in a measure blind to other facts. He now not only gives to his favorite facts an undue importance, but his imagination invests them with hues that do not belong to them ([13], pp. 33–4).

One of Hooker's chief examples was Bishop George Berkeley. Hooker judged Berkeley among the great minds of the eighteenth century, yet found that he committed elementary inductive fallacies in his enthusiasm for the virtues of Tar Water.[10] Berkeley's great mistake coincided with the one Hooker found most prevalent in medicine: the fallacy of *post hoc, ergo propter hoc* ([13], pp. 7–8, 15). Hooker describes this form of fallacious reasoning as "*the too ready disposition to consider whatever follows a cause as being the result of that cause*" ([13], p. 8). The prime instance is that when medicines are administered and a cure follows, the medicine is readily accepted as the cure, despite many possible alternative explanations. New remedies and measures get adopted incautiously and without proper observation and testing in medicine ([15], pp. 22–8). Hooker accuses many figures in the history of science and the history of medicine of similar elementary fallacies of thinking, owing to their specific enthusiasm, their following of ideology, and their conformity to fads and fashions. Not surprisingly, he judges this fallacy "the great source of quackery" ([12], pp. 65, 71, 80, 87; [16], p. 55).

What these physicians almost always forgot, in Hooker's estimation, is twofold. First, they forgot that the curative power of nature itself is a better explanation in these contexts than the particular substance or procedure used to explain the cure or elimination of disease. Each enthusiast fell victim to the great human temptation to be a "one idea man" by becoming disposed "to adopt exclusive views and notions," blinding the would-be investigator to other facts and possibilities ([13], pp. 34–5). Hooker argued that even well informed patients and physicians are at times blinded by the hope of a remedy, although a prudent and distanced judgment suggests its inefficacy.

Second, Hooker thought the quack and all who searched single-mindedly for remedies failed to appreciate how much uncertainty there is in medicine, owing to complex causal relationships. So multifarious are these phenomena that they "prevent uniformity in the effects of remedies." Hooker insists that organs affect other organs in complex and unpredictable fashion, that there are many unseen causes or agents, that mental influences sometimes play a role, and that there are many idiosyncrasies, or individual peculiarities in medicine ([12], Chapter 1, especially p. 27).

What, one may ask, have these arguments about delusions and fallacies to do with medical ethics? Although Hooker leaves the precise connection unclear, there is at least the indirect connection that he looked to the *Code of Ethics* of the AMA as a means of exerting moral leadership and control over a profession vulnerable to these delusions. More importantly, Hooker thought quackery, especially Thompsonism, "an unmitigated evil" ([12], p. 45), and he thought there were many "tendencies of an evil character" at work in the medicine of his period ([15], pp. 22, 28). His point appears to be both that many disastrous errors made by physicians were avoidable and that many prescribed medicines have been so destructive of life and health that it is unconscionable to continue with them.

Hooker clearly believed that some forms of these prevailing abuses and delusions involved "bad conduct" by "dishonorable men" ([13], p. 86). However, Hooker recognized that many quacks and many ignorant or gullible physicians were, like Berkeley, well intentioned, and so not dishonest or morally disreputable. When he criticizes disreputable conduct, he does not always take the persons criticized to be mischievous, although he also thinks that most delusions rest on an eliminable ignorance that ought to have been eliminated. Hooker's harshest criticisms concerning eliminable ignorance were reserved for homeopaths [14]. The followers of Samuel Hahnemann, he argued, were deficient in their scientific acumen. They indefensibly undervalued medicine as an observational science. So weak was their account of confirmation that "anything may be made to prove anything that may be desired." This led to certain forms of dishonorable conduct and at the same time provided grounds for excluding homeopaths from the category of *physicians* ([12], p. 136; [13], pp. 86, 24f; [15], p. 15; [14], p. ix).

Hooker also raises, in a subtle but underexplored fashion, one of the more vexing questions in medical ethics today: the conditions under which there is an obligation to disclose incompetence or unscrupulous behavior in colleagues and others encountered in professional life. Such disclosures seem essential in order to preserve trust with the public and with professional colleagues (as the *AMA's Principles of Medical Ethics* today explicitly maintain). Yet exposés were then, and still are, uncommon as a result of bonds of loyalty, which were accented in the Hippocratic and gentlemanly traditions of medical ethics promoted by Percival.

Hooker held that in the face of bona fide misrepresentation, fraud, unethical conduct, or incompetence, the physician has an obligation to confront the problem and to encourage the repudiation of improper activities. In some cases there may be an obligation to take specific action to correct inappropriate behavior or to confront any wall of silence in the profession.

CONCLUSION

Hooker was the first physician in the history of medical ethics to defend something like an autonomy model, but he did not espouse a pure autonomy model. Hooker, no more recommends respect for autonomy for the sake of autonomy than does the AMA's *Code*. Hooker's concerns were with expediency in disclosure and truthtelling, rather than with the promotion of autonomous decision-making or informed consent. Yet the latter have become central to the contemporary vision of an autonomy model. Hooker's failure to bring more autonomy into his medical ethics is not surprising, inasmuch as the nineteenth-century social context was not rights-oriented, and practices of disclosure to patients were commonly conceived in terms of therapeutic benefit rather than individual rights. The idea that patients should be enabled to understand their situation so that they could participate in a dialogue with physicians was an idea whose time was yet to come.

NOTES

[1] Thomas Gisborne a century earlier had held similar views.

[2] Hooker's major works all deal with these problems. See also, for the related ethical and organizational implications, [17], Chapters. 2–3, and [7], pp. 31–6.

[3] American physicians had been out to regulate their fellows by the erection of professional standards at least since a set of influential moral rules modeled on Percival and published by Boston physicians in 1808 as *Boston Medical Police* ([23], see also: [17], p. 2; [3], p. 302, and [4].

[4] The chairman of the AMA's drafting committee for the Code, Isaac Hays, wrote a note accompanying the committee report: "On examining a great number of codes of ethics adopted by different societies in the United States, it was found that they were all based on that by Dr. Percival, and that the phrases of this writer were preserved to a considerable extent in all of them." He also noted that some of the sections in the new code were "in the words of the late Dr. Rush" [10]. See [7], pp. 35–6; [8], p. 5. Partially as an attempt to persuade the New York Society to reenact the national code, Austin Flint [8] provided commentary on the code that shines as the most carefully reasoned work on medical ethics since Hooker.

The dominance of Percival was everywhere evident even as late as Flint's commentary. For example, considerable feeling for the pervasive sense of the code's adequacy is found D. W. Cathell's *The Physician Himself* (1882).

> Dr. Thomas Percival['s code] . . . until now has governed our whole profession throughout this broad land . . . stands like a lighthouse to guide and direct all who wish to sail in an honorable course. . . .
>
> By its justness this code remains as fresh and beautiful to-day as when Percival penned it seventy-five years ago (([5], pp. 42–3).

Nathan Davis later noted that the AMA's *Code* was "copied" chiefly from Percival, although the literal wording was often recast to fit a broader medical context more appropriate to mid-nineteenth-century America ([6], pp. 35–40. Davis was Dean and Professor of Medicine at Northwestern University, a politically active member of the AMA, and historian of American Medicine. He personally witnessed developments in the *Code* for over 50 years. The historical view that he took, in lectures from 1892–1897, was that "It was not until the end of the eighteenth century that the Hippocratic Code was more fully *discussed, revised, and extended* by Sir Thomas Percival" and became the living code of the AMA through its conventions, which simply lifted material from Percival, Rush, and Gregory ([6], pp. 189–91, *italics* added).

[5] A widely cited edition is Chauncey D. Leake's *Percival's Medical Ethics* [18], but it is not entirely reliable. The full title placed by Percival on this "Note" (dropped by Leake, in part) is, "A Physician Should Be Minister of Hope and Comfort to the Sick. – Enquiry, how far it is justifiable to violate Truth for the Supposed Benefit of the patient" ([21], p. xv).

[6] Gisborne says his critique is directed at "Dr. Percival's *Medical Jurisprudence* [20], p. 15."

[7] Hooker discusses this passage in Hutcheson, and Percival's use of it in [12], p. 379.
[8] "In truth, expediency and right always correspond, and would be seen to do so, if we could always see the end from the beginning" ([12], p. 360).
[9] [12], pp. 376–7; but compare 185–6.
[10] [13], pp. 12–6, 20, 22, 27, 31. Berkeley's errors are also briefly discussed by Hooker in [12], p. 198. Hooker's sources were Berkeley's *Siris: . . . concerning the Virtues of Tar Water*, 2nd ed. of 1747, and his essay, "Farther Thoughts on Tar Water."

BIBLIOGRAPHY

1. American Medical Association: 1847, "Code of Ethics," *Minutes of the Proceedings of the National Medical Convention held in the City of Philadelphia, in May 1847*, pp. 83–106; this volume, pp. 65–88.
2. Bell, J.: 1847, "Introduction to AMA *Code of Ethics*"; this volume pp. 65–72.
3. Burns, C.: 1977, "Reciprocity in the Development of Anglo-American Medical Ethics, 1765–1865," in Burns, C. (ed.), Legacies in Ethics and Medicine, pp. 300–6, Science History Publications, New York; this volume, pp. 135–144.
4. Burns, C.: 1978, "Medical Ethics, History of: North America: Seventeenth to Nineteenth Century," in Warren Reich (ed.), Encyclopedia of Bioethics, Vol. III, pp. 963–5, The Free Press, New York.
5. Cathell, D.: 1882, The Physician Himself, 2nd ed., (reprint) Arno Press, The New York Times, 1972, New York.
6. Davis, N.: 1903, History of Medicine, with the Code of Medical Ethics, Cleveland Press, Chicago.
7. Fishbein, M.: 1947, A History of the American Medical Association 1847 to 1947, W. B. Saunders Co., Philadelphia.
8. Flint, A.: 1895, Medical Ethics and Etiquette: The Code of Ethics Adopted by the American Medical Association, with Commentaries, D. Appleton, New York.
9. Gisborne, T.: 1794, An Enquiry into the Duties of Men in the Higher and Middle Classes of Society in Great Britain Resulting from their Respective Stations, Professions and Employment, B. and J. White, London.
10. Hays, I.: 1847, "Note to AMA *Code of Ethics*," this volume, pp. 73–75.
11. Hooker, W.: 1844, Dissertation, on the Respect Due to the Medical Profession, and the Reasons that It is not Awarded By the Community, J. G. Cooley, Norwich, CT.
12. Hooker, W.: 1849, Physician and Patient; or, a Practical View of the Mutual Duties, Relations and Interests of the Medical Profession and the Community, Baker and Scribner, New York.
13. Hooker, W.: 1850, Lessons from the History of Medical Delusions, Baker & Scribner, New York.
14. Hooker, W.: 1852, Homœopathy: An Examination of its Doctrines and Evidences, Charles Scribner, New York.
15. Hooker, W.: 1852, Inaugural Address: The Present Mental Attitude and Tendencies of the

Medical Profession, T. J. Stafford, New Haven, CT.
16. Hooker, W.: 1857, Rational Therapeutics; or the Comparative Value of Different Curative Means, and the Principles of Their Application, John Wilson and Son, Boston.
17. Konold, D.: 1962, A History of American Medical Ethics 1847–1912, State Historical Society of Wisconsin, Madison, WI.
18. Leake, C.: 1975, Percival's Medical Ethics, Robert E. Kreiger Publishing Co., New York.
19. Pellegrino, E.: 1986, "Percival's *Medical Ethics*: The Moral Philosophy of an 18th-Century English Gentleman," *Archives of Internal Medicine* 146, 2265–9.
20. Percival, T.: 1794, *Medical Jurisprudence or a Code of Ethics and Institutes Adopted to the Professions of Physic and Surgery*, privately circulated, Manchester, UK.
21. Percival, T.: 1803, *Medical Ethics; Or, A Code of Institutes and Precepts, Adapted to the Professional Conduct of Physicians and Surgeons*, J. Johnson, London.
22. Sidgwick, H.: 1981, *The Methods of Ethics*, 7th ed., Hackett Publishing Co., Indianapolis, IN.
23. Warren, J., Hayward, L., and Fleet, J.: 1808, *The Boston Medical Police*, Association of Boston Physicians, Boston, this volume, pp. 41–46.

CHAPTER 5

ROBERT M. VEATCH

DIVERGING TRADITIONS: PROFESSIONAL AND RELIGIOUS MEDICAL ETHICS OF THE NINETEENTH CENTURY

The existing literature provides only a poor understanding of the history of nineteenth-century American medical ethics. That literature gives us considerable insight into the background of the passage of the American Medical Association's *Code of Ethics* of 1847, including the widely adopted state and local association codes beginning with the Boston Medical Association code of 1808 ([3], [4], [22], [24], [25]). What we do not know much about is how these codifications played among physicians and non-physician intellectuals concerned about the morality of the practice of medicine in a period of considerable controversy and confusion. Of particular interest is the reception, if any, of the 1847 AMA *Code* and its predecessors among those doing what we would now call medical ethics in the religious traditions of the time. While modern associations of physicians have taken considerable interest in the writing of codes of ethics to govern the relations of physicians and patients, it would be a serious mistake to assume that medical ethics is synonymous with these professionally generated codes. Not only do physicians as well as other health professionals have views on the ethics of the lay-professional relation that may differ from the codes of their organized associations, many other groups (see [15]) – religious, governmental, and philosophical – have had well-developed positions on the ethics of medicine and the roles of the patient and professional healer.

Of particular interest is the relation, if any, of the religious communities to the AMA code and other professional activities to articulate a medical ethic. This essay explores that relation – or, as we shall see – the lack thereof. The thesis of the essay is that independent medical ethical traditions existed in the religious communities of the nineteenth century United States, traditions that

R. Baker (ed.), The Codification of Medical Morality, 121–132.
© 1995 *Kluwer Academic Publishers. Printed in the Netherlands.*

were oblivious to, or perhaps intentionally indifferent to, the codification promulgated by organized medicine.

THE EXISTENCE OF RELIGIOUS TRADITIONS IN MEDICAL ETHICS

It is widely assumed by students of medical ethics that the Hippocratic tradition has been the sole or dominant view at least since the Christianization of the ancient world. Some scholars have hypothesized that, although there were clearly competing schools of Greek medicine – including medical ethics – there was a convergence of Christian thought with Hippocratic medical ethics soon after Constantine that led to the dominance of Hippocratic ethics ([7], pp. 66, 159; [8], p. 62, n. 45; [36], pp. 157, 170). Carol Mason Spicer and I have examined this hypothesis and have found it wanting. Not only are there important substantive differences between the religious ethical traditions and Hippocratic medicine, there is almost no evidence of contact between them, at least during the first eight centuries of the Christian era. We could find only two explicit references in the church fathers to Hippocratic ethical writings (to the fourth century church fathers, Jerome and Gregory of Nazianzus [35]). Both of these consciously distinguish Hippocratic and Christian medicine. From about the eighth to the twelfth centuries there was a much more complex intermingling of religious and medical roles including the existence of a Christianized version of the Oath with earliest manuscripts dating from the tenth century entitled "Oath According to Hippocrates in so far as a Christian May Swear It," [18], which is sometimes taken as evidence for convergence, but can at least as well be taken as evidence that medieval Christian writers were unable to accept many provisions of the Hippocratic writings.

By about the twelfth century, there was the beginning of a secularization and professionalization of medicine. Priests were forbidden to practice medicine ([6], p. 51). It can be argued that the enlightenment simply brings the final stages of secularization and professionalization of medicine. McCullough [28] has suggested that with Percival's Manchester code, written in the 1790s and published in 1803, we have a "radical shift" from Gregory's approach and a "major shift in kind" in Anglo-American medical ethics. There is less clear connection between the dominant religious/philosophical scholarship of the day (including that of Hutcheson and Hume) and a more isolated, independent concern with intra-professional matters of physician authority and power.

Kelly [21] claims that secularization and professionalization of medical ethics among organized medical professionals generated a backlash among religious scholars who perceived a greater need for an explicitly religious moral framework to differentiate their positions from the matters of concern to medical professionals.

While he was writing with specific reference to the development of Catholic medical ethics in North America, the same point could be made with regard to Judaism and Protestantism as well. All three groups have long had ethical traditions with at least implicit medical ethical implications.

Judaism finds the roots of its medical ethics in the Talmud and commentaries. Jakobovits [16] points out that Judaism never even had a Jewish version of the Hippocratic Oath, relying instead on its own long medical, ethical heritage including the Oath of Asaph, the writings of Jewish rabbi-physicians, and more recent medical ethical documents such as the eighteenth-century prayer attributed to Maimonides. While there is no specific research done as yet on Jewish treatment of medical ethical issues in the nineteenth century, there is no evidence that Jewish scholarship took cognizance of the AMA *Code* at mid-century. While Talmudic scholarship has shown a respect for secular work in medicine, it would be totally out of keeping with this tradition of scholarship to credit the consensus of a group of primarily gentile physicians meeting in Philadelphia in May of 1847 with insights worthy of attention to rabbinical scholarship.

Likewise, Protestant thought in the mid-nineteenth century showed no concern about the ethics activities of the AMA. It was dominated primarily by other, more timely matters: first the Great Revival of 1830 ([5], [10], [38]) and then the voluntary charitable societies and abolition movement that followed [2], [9], [14]. The dominant theme related to medical ethics was the emphasis on diet, temperance, and simple, natural remedies. This continued the significant contribution of John Wesley as seen in his phenomenally successful and influential *Primitive Physics* [39].

The influence of Wesley particularly manifest itself in the nineteenth-century American movements of sectarian Protestantism. Mormons, Seventh Day Adventists, Jehovah's Witnesses, and Christian Science are all mid- to late-nineteenth-century sectarian movements with significant medical components emphasizing the link between disease, on the one hand, and diet and life-style on the other. Although all but Christian Science make use of orthodox medical knowledge, their unique doctrines relating healing to their religious beliefs make them less interested in the authority of the American Medical Association on matters moral. They all use specialized healers or practitioners and, in varying degrees, incorporate moral positions that would be incomprehensible to those outside the faith. For them the source of moral authority and knowledge was within their sectarian communities, not in the AMA ([11], [17]).

The detailed histories of the medical ethics of Jewish and Protestant groups in the nineteenth history cannot be developed here. Rather, as a way of exploring my hypothesis in detail, I shall examine Roman Catholic moral theology and its treatment of what we would call "medical ethics."

ROMAN CATHOLIC MEDICAL MORALITY

The Roman Catholic tradition in the United States increasingly differentiated itself from the medical ethics of organized medicine during the nineteenth century. It is not that there was an overt, hostile reaction to the development of local, state, and national codes of ethics such as those of the AMA. Rather the methodology and substantive normative ethical concerns of the Catholic theologians and physicians writing on the subject simply took them in a significantly different direction. As far as I can tell there was no public response or even acknowledged awareness of the AMA's adoption of its code in 1847. There was, however, a rich tradition of continuing pursuit of the morality of the physician's role. At the beginning of the century, this was primarily based on use of Catholic materials from the European continent. By the end of the century, American materials clearly in the same tradition were common.

There is a perception, at least by later commentators, that with the increasing professionalization of medicine, the concern of organized Anglo-American medicine turned to problems of power, authority, and particularly relations among medical professionals and their competitors [21]. It has even been suggested that Catholic commentators refused to use the term "medical ethics" for their work for fear of confusing "real" morality with the questions of intra-professional etiquette being addressed by professionals, but I find no nineteenth century evidence that Catholic commentators used the language the way they did for this reason; the term "moral theology" had been in use for years.

Three closely related genre of Catholic moral literature spoke to issues of medical ethics during the nineteenth century [21]. First were the moral theology manuals. They have their origins in the seventeenth century, but the nineteenth century works evolved from the 1785 expanded edition of Alfonso Liguori's *Theologia moralis* [27]. Works throughout the nineteenth century following this model are described as "nearly carbon copies of their predecessors" ([21], p. 30; see, for example, [13], [22], [24], [31], [32]). None of these is exclusively a medico-moral work, but certain important issues for medicine were covered.

A second group of writings approached medical ethical issues under the rubric of *casus conscientiae* or cases of conscience ([12], [26], [37]). They followed the same organizational structure as the manuals of moral theology, but claim to base their conclusions solely on "natural" human reason.

Finally, an important genre of Catholic writing in the nineteenth century was the volumes on what was called "pastoral medicine." They were designed to serve two purposes: to provide medical knowledge for pastors and theological and ethical preparation for medical practitioners. Carl Capellmann's *Pastoral Medicine* [6] is the first to appear in English at the time when Catholic moral theology was just beginning to be written in the vernacular. It appeared in English one year after the original publication and explicitly acknowledges its dependence on the moral theology works of Gury, Liguori, and Scavini ([6], see Table of Contents page).

In an extensive search of these documents, it became quite clear that they operate in a different world from the professional medical ethical literature of the time. It would be understandable if some of the Europeans were not in conversation with the Anglo-American professional ethical literature, but the problem is the same for the American authors. They are working in an ethical tradition that is not in communication with the medical professional organizations. Still they have positioned themselves to provide authoritative advice for physicians and patients – at least those within the Catholic tradition.

A SUBSTANTIVE COMPARISON OF THE TRADITIONS

The significance of the existence of multiple medical ethical systems, each ignorant of or indifferent to the existence of the other, remains to be explored. It would seem that this would be a matter of concern for Catholic physicians who are simultaneously loyal members of the AMA or for Catholic patients who are obtaining their health care from physicians who are guided by AMA ethics, but not Catholic. A similar concern would be plausible for Jews, Seventh Day Adventists, and others standing in some specific religious medico-moral tradition who are subject to the AMA perspective, either as physician-members, or as patients in the care of an AMA physician.

The critical question is: How different are the religious and professional medical ethical frameworks ? We will look briefly at their methodologies and their normative concerns.

Medico-Moral Methodologies

We know quite clearly how the AMA went about writing its *Code of Ethics*. The AMA committee writing the draft took whole sections verbatim from

Percival, either directly or mediated through intermediary documents, and incorporated material from Benjamin Rush as well, as had been done in earlier state and local codes such as those Boston in 1808, New York in 1823, and Philadelphia in 1843 ([3], [4]). The working assumption was that a profession is responsible for articulating its own code of ethics. It drew on other medical professional writing, but there is no evidence of any interest in the major philosophical or theological schools of thought of the day.

John Bell's opening sentence in his Introduction to the 1847 *Code* claims that medical ethics (the term used by the physicians) "as a branch of general ethics, must rest on the basis of religion and morality" [1], but that is the only reference to the general disciplines of ethics, morality, or religion. The project is clearly one belonging to the profession, not to the theologians, philosophers, or the general public. The AMA's professional medical ethics is detached from the foundations of ethics, whether secular or religious.

By contrast the Catholic medico-moral literature we have examined clearly sees any positions on the ethics in the medical role to be derivative from a more general moral theology. Any claim of moral authority by a medical professional body is appropriately viewed with skepticism. In the early moral theology manuals, the organizational structure is primarily around the Biblical ten commandments and the sacraments ([21], pp. 24, 30). An alternative organizational structure is around the classical virtues ([21], p. 38). Some have a special section on the obligations of medical personnel. Regardless, there is a working assumption that there is a general framework for doing moral theology. It includes presuppositions about methods of justification and sources of authority. Once that framework is in place, the implications for the medical roles follow. Thus most questions of interest to medical analysis arise under the rubric of the fifth or sixth commandments (the commandments against killing and committing adultery). Under the heading of the fifth commandment abortion, euthanasia, suicide, castration, and mutilation are treated ([21], p. 31).

Under the sixth commandment come the titillating issues of fornication, rape, adultery, incest, coital positions, contraception, homosexuality, and masturbation. Additional questions arise under the heading of the sacraments, particularly matrimony, under which some of the sixth commandment issues sometimes are covered.

Normative Ethical Issues

The differences in moral methodology thus lead to important differences in substance between the AMA and other professionally articulated medical ethical codes and the medico-moral framework of the theologians.

It is striking that the substantive issues developed ad nauseam in the Catholic literature scarcely get mentioned in the professional codes. Jonsen and Hellegers [19] have argued that the professional codes emphasize development of the virtues rather than the norms of right conduct (duties or obligations) that are the focus of the Catholic medico-moral literature. There is some reason to doubt this claim given the language of duties that occurs throughout the 1847 code [30]. Still the duties of the physician to the patient are expressed in extremely vague and general terms: physicians are to "minister to the sick with due impressions of the importance of their office" ([1], Chapter One, Article 1, § 1). Much of the emphasis is indeed on the character of the physician including the oft-quoted, controversial virtues of the gentleman: tenderness, firmness, condescension, and authority ([1], Chapter One, Article 1, § 1). Even as virtues they are strangely at odds with the cardinal virtues that provide the structure for the virtue manuals in Catholic moral theology: wisdom, temperance, courage, and justice.

Even though the 1847 AMA code does include some talk of duties, the handling of the duties differs from the Catholic medico-moral tradition of the nineteenth century as much as the treatment of the virtues does. The contrast can be summarized by saying that none of the dominant duty themes of one tradition are comparable to those of the other. We can see the difference by summarizing the main themes of the Catholic literature.

As we have noted, the main structure of the Catholic medico-moral discussion is often around the fifth and sixth commandments. Capellmann follows this standard approach, devoting the first ten pages of his work to abortion and "perforation of the living fetus" ([6], pp. 10–20). It comes as no surprise that the Catholic literature gives substantial attention to this issue. In fact, through the nineteenth century the Catholic concern about abortion actually heightened, leading to Pope Pius IX's 1869 Constitution *Apostolicae Sedis*, which eliminated any lingering doubt about the moral difference between formed and unformed fetuses and made excommunication the penalty for abortion ([29], pp. 24–31). By contrast the AMA is silent on the subject of abortion.

Immediately following the treatment of abortion in Capellmann's work is an equally detailed nine pages on "Operations attended with risk to life" ([6], pp. 20–8). Here is a detailed discussion leading to the conclusion that "no one is obliged to undergo a severe operation involving risk of life, although affording, at the same time, a hope of its preservation" ([6], p. 29). In fact, Capellmann makes a point that excessively risky operations are morally forbidden.

The AMA treatment of this subject is much more shallow. The focus seems to be on the possibility that a physician would abandon a patient in a hopeless condition. Patients should not be abandoned, we are told, because, if the physician stays with the patient, pain and mental anguish may be relieved. There is no awareness of the possibility that the burden to the patient of treatment may be overwhelming and provide a moral justification for forgoing further care.

Capellmann's discussion of the fifth commandment then includes a long discussion of morphine, chloroform, and animal magnetism, all dealing with the issue of whether these are so dangerous that they are morally prohibited. His conclusion is that "The physician should always make use of such remedies as are regarded *safe* in the existing state of medical science" ([6], p. 29). Nothing remotely similar occurs in the AMA Code.

The even longer discussion of the sixth commandment occupies over forty pages covering such topics as masturbation, "pollutions" (nocturnal emissions), and the uses of marriage. The latter covers in great detail the questions of ethical and unethical copulation, contraception, and coitus interruptus. William Dassel, the American priest who translated Capellmann, obviously struggling with these delicate subjects, explains in his preface that although he favors the use of the vernacular, he has attempted "to lessen the disgust necessarily provoked by unavoidable details, but putting them into a Latin disguise. . . ." ([6], p. iv).

By mid-century, the pope had put to rest any doubt about the Catholic view on contraception. On May 21, 1851, the Holy Office issued a decree that states:

> The Apostolic See is asked what theological note is to be applied to the following propositions: (1) a married couple may practice contraception for morally good motives; (2) this form of marital intercourse is not certainly against the natural law. The Holy Office answers: the first proposition is scandalous, erroneous and contrary to the natural law of marriage; the second is scandalous and implicitly condemned in proposition 49 of Innocent XI (cited in [20]).

By contrast, the AMA in 1847 makes no mention of contraception and related ethical problems of marital relations.

Two additional normative themes are worth mentioning because of the contrast between the Catholic and AMA positions regarding them. First, Catholic moral theology has long emphasized, in cases of terminal illness, the need to disclose to the patient his diagnosis. As early as the fifteenth century, Antonius of Florence (1477) taught of the necessity of the physician to warn patients of their impending death so that they might adequately prepare their souls ([21], p. 26). The same theme appears in the nineteenth-century manuals and pastoral medicine texts ([6], pp. 167–9).

The AMA in 1847 provides a much more Hippocratic, paternalistic reading of the physician's duty. He "should not be forward to make gloomy prognostications, because they savor of empiricism, by magnifying the importance of his services in the treatment or cure of the disease. But he should not fail, on proper occasions, to give to the friends of the patient timely notice of danger, when it really occurs; and even to the patient himself, if absolutely necessary" ([1], Chapter One, Article 1, § 4). The conflict with the traditional duty of confidentiality appears overlooked. Certainly, there is no awareness of the possibility that the patient may need this information to make preparation – secular or religious – for his death.

Finally, there is a difference, at least in emphasis, in what might be called the social ethics of medicine. The AMA in 1847 departs from Hippocratic tradition in including an explicitly social dimension. The third chapter deals with "duties of the profession to the public, and of the obligations of the public to the profession." This, however, deals with newly emerging matters of public health – "medical policy, public hygiene, and legal medicine" – and with the duty of the physician in an epidemic rather than questions we would describe as the right of the poor to access to a physician.

By contrast Catholic moral theology has long emphasized a social ethic for medicine that includes a duty to treat the poor without fee. Kelly traces this Catholic medico-moral theme back as far as Antonius of Florence in the fifteenth century ([21], p. 26).

Thus it seems clear that both in moral methodology and in substantive normative issues, the tradition of Catholic moral theology and that of organized professional medicine in the United States were operating in different worlds. Their sources of authority were different; the issues they were worried about were different; and even their concept of ethics was different. It is understandable why the AMA might not address itself to the Catholic agenda; it is less clear why the Catholic writers felt comfortable ignoring the AMA.

A CONCLUDING PUZZLE

This suggests a final question that I have not been able to answer. During the nineteenth century in American medical ethics positions were crystallizing. The AMA had to deal with what we now call nonorthodox practitioners. Much of the AMA's energy was devoted to clarifying how physicians should relate to the nonorthodox healers [23]. It seems to have formulated a code of ethics in part to convey that medicine was a profession with autonomy in matters of ethics. In doing so, however, the AMA took stands on some matters that should have made those standing in the Catholic tradition uncomfortable – on disclosure to patients and on the source of authority in ethics, for instance. More importantly, it did not address what to those in the Catholic tradition was central – abortion, contraception, mutilation, the care of the dying, and social responsibility for the poor.

How can it be that Catholic physicians were not in a terrible crisis, caught between two competing claims on them for loyalty? How can it be that Catholic lay persons were not equally troubled, worried that they would get medical care from a physician who loyally subscribed to the new code of the AMA and submitted himself to the AMA's authority on questions that could easily have been perceived as matters only resolvable (for Catholics) through the tradition of moral theology?

Unless I have missed something important, the two traditions simply were not in communication throughout the nineteenth century. Was it that they simply perceived no conflict – a hypothesis that seems incredible given the obvious disagreements especially on matters of authority? Or was it that each really did not know what the other group was doing – a hypothesis equally incredible given the visibility of each of the traditions? I must leave this puzzle for solving at some later time. What seems clear at this point is that there were separate medical ethical traditions in the nineteenth century, traditions apparently oblivious to the methods and conclusions of others, traditions that did not converse with one another.

BIBLIOGRAPHY

1. American Medical Association: 1848, *Code of Medical Ethics: Adopted by the American Medical Association at Philadelphia, May, 1847, and by the New York Academy of Medicine in October, 1847*, H. Ludwig and Company, New York; this volume, pp. 65–88.
2. Barnes, G.: 1933, *The Anti-Slavery Impulse: 1830–1844*, (1933 reissue) Harcourt, Brace & World, New York.
3. Burns, C.: 1977, "Reciprocity in the Development of Anglo-American Medical Ethics, 1765–1865," in Burns, C. (ed.), *Legacies in Ethics and Medicine*, pp. 300–6, Science History Publications, New York; this volume, pp. 135–144.
4. Burns, C.: 1978, "Medical Ethics, History of: North America: Seventeenth to Nineteenth Century," in Reich, W. (ed.) *Encyclopedia of Bioethics*, Vol. III, pp. 963–5, The Free Press, New York.
5. Bushnell, H.: 1861, *Christian Nurture*, (1962 reissue) Yale University Press, New Haven, CT.
6. Capellmann, C.: 1879, *Pastoral Medicine*, (W. Dassel, trans.), Pustet, New York.
7. Carrick, P.: 1985, *Medical Ethics in Antiquity: Philosophical Perspectives on Abortion and Euthanasia*, D. Reidel Publishing Company, Dordrecht.
8. Edelstein, L.: 1967, "The Hippocratic Oath: Text, Translation and Interpretation," in Temkin, O. and Temkin, C. (eds.), *Ancient Medicine: Selected Papers of Ludwig Edelstein*, pp.

3–63, The Johns Hopkins Press, Baltimore, MD.

9. Filler, L.: 1960, *The Crusade Against Slavery: 1830–1860*, Harper & Row, New York.
10. Finney, C.: 1835, *Revivals of Religion* (1962 reissue), The Moody Bible Institute, Chicago.
11. Fuller, R.: 1989, *Alternative Medicine and American Religious Life*, Oxford University Press, Oxford.
12. Gury, J.: 1866, *Casus conscientiae in praecipuas guestiones theologiae moralis*, 6th ed. (1881), Briday, Lyons.
13. Gury, J.: 1869, *Compendium theologiae moralis*, 18th ed., H. Pelaguad, Paris.
14. Hobbs, G.: 1964, *The Crusade Against Slavery: 1830*, Harcourt, Brace & World, New York.
15. Hooker, W.: 1849, *Physician and Patient; or, a Practical View of the Mutual Duties, Relations and Interests of the Medical Profession and the Community*, Baker and Scribner, New York.
16. Jakobovits, I.: 1978, "Judaism," in Reich, W. (ed.), *Encyclopedia of Bioethics*, Vol. 2, pp. 791–802, The Free Press, New York.
17. Jones, R. K.: 1985, *Sickness and Sectarianism*, Gower, Hampshire, England.
18. Jones, W.: 1924, *The Doctor's Oath: An Essay In The History Of Medicine*, Cambridge University Press, Cambridge.
19. Jonsen, A. and Hellegers, A.: 1974, "Conceptual Foundations for an Ethics of Medical Care," in Tancredi, L. (ed.), *Ethics of Health Care*, pp. 3–20, National Academy of Sciences, Washington, DC.
20. Jonsen, A. and Toulmin, S.: 1988, *The Abuse of Casuistry: A History of Moral Reasoning*, University of California Press, Berkeley.
21. Kelly, D.: 1979, *The Emergence of Roman Catholic Medical Ethics in North America: An Historical – Methodological – Bibliographical Study*, Edwin Mellen Press, New York.
22. Kendrick, F.: 1860–61, *Theologia moralis*, H. Dessain, Mechlin.
23. King, L.: 1982, "The 'Old Code' of Medical Ethics and Some Problems It had to Face," *Journal of the American Medical Association* 248, 2329–33.
24. Konings, A.: 1880, *Theologia moralis novissimi ecclesiae doctoris S. Alphonsi, in compendium redacta, et usui venerabilis cleri americana accomodata*, 4th ed., Benziger Brothers, New York.
25. Konold, D.: 1962, *A History of American Medical Ethics 1847–1912*, State Historical Society of Wisconsin, Madison, WI.
26. Lehmkuhl, A.: 1907, *Casus conscientiae ad usum confessariorum compositi et soluti*, 3rd ed., Herder, Fribourg.
27. Liguori, A. M. De': 1748–85, *Theologia moralis* (1905–1912, Editio nova), L. Gaude (ed.), 4 Vols., Vaticana, Roma.
28. McCullough, L.: 1985, "Virtue, Etiquette, and Anglo-American Medical Ethics in the Eighteenth and Nineteenth Centuries," in Shelp, E. (ed.) *Virtue and Medicine: Exploration in the Character of Medicine*, pp. 81–92, Reidel, Dordrecht.
29. Pius IX: 1923, "const. *Apostolicae Sedis*, 12 Oct. 1869," *Codicis Iuris Canonici Fontes*, P. Gasparri (ed.), Typis Polyglottis Vaticanis, Rome, Vol. 3.
30. Ramsey, P.: 1974, "Commentary: Jonsen and Hellegers," in Tancredi, L. (ed.), *Ethics of Health Care*, pp. 3–20, National Academy of Sciences, Washington DC.
31. Sabetti, A.: 1889, *Compendium theologiae moralis a Joanne Gury, S. J. primo exaratum et*

deinde ab Antio Ballerini . . . auctum nunc vero ad breviorem formam redactum atque ad usum seminariorum hujas reegionis accomodatum, 4th ed., F. Pustet, New York.
32. Scavini, P.: 1850, *Theologia moralis universa ad menetem S. Alphonsi M. de Ligorio*, 8th ed., Ernesto Oliva, Milan.
33. Scotti, A.: 1836, *Catechismo medico ossio suiluppo delle dottrine cotta che conciliano la religione cotta medicina de piu nuovi capi cresciuto nella presente edizione dal ch. suo autore*, Facolta.
34. Vanderpool, H., "John Wesley's Medicine for the Masses," (unpublished paper).
35. Veatch, R. and Mason, C.: 1987, "Hippocratic vs. Judeo-Christian Medical Ethics: Principles in Conflict," *The Journal of Religious Ethics* 15, 86–105.
36. Verhey, A.: 1984, "The Doctor's Oath – and a Christian Swearing It," in Smith, D. (ed.), *Respect and Care in Medical Ethic*, pp. 157, 170, University Press of America, Lanham, MD.
37. V[illada], P.: 1836, *Casus conscientiae his praesertim temporibus accomodati, propositi, ac resoultl*, Alfred, Vromant, Brussels.
38. Walker, W.: 1918, *A History of the Christian Church* (1959 reissue), Charles Scribner's Sons, New York.
39. Wesley, J.: 1785, *Primitive Physics*, 21st ed., J Paramore, London.

PART TWO

MEDICAL ETHICS AND MEDICAL JURISPRUDENCE IN NINETEENTH-CENTURY BRITAIN

CHAPTER 6

CHESTER BURNS

RECIPROCITY IN THE DEVELOPMENT OF ANGLO-AMERICAN MEDICAL ETHICS, 1765–1865

From the beginnings of recorded human history, ideals and ideas about values have been associated with the personal and professional activities of medical practitioners. The professional values can be classified best under the following four headings: (1) the education of medical practitioners, (2) consultations with other practitioners, (3) transactions between physicians and patients, and (4) relationships between medical practitioners and communities. During the one hundred years encompassed by this study, Anglo-American physicians experienced value changes within all four of these categories of interpersonal relationships (see [3], [8]). These changes are well illustrated by the ideals of three British physicians: John Gregory (1724–1773), Thomas Percival (1740–1804), and Michael Ryan (1800–1841).

After joining the Edinburgh faculty as professor of medicine in 1765, Gregory gave several introductory lectures about the qualifications and duties of physicians. He published six of them in 1772 [4]. His ideals primarily involved the education of a physician and the nature of medical science. He strongly believed that a formal education in particular subjects constituted the ethical basis for medical practice. Education, science, and ethics were inseparable. Even when dealing with the public, Gregory's scientific emphasis continued. He was less concerned about what physicians should do for the community than about what laymen scientists could do for medicine. Somewhat randomly he exhorted his students to attend to the professional decorums that underlay interactions of practitioners and transactions between practitioners and patients. A more detailed analysis of these latter two groups of professional ideals was made by Thomas Percival, a Manchester physician who had carefully studied Gregory's book.

R. Baker (ed.), The Codification of Medical Morality, 135–143.
© 1995 *Kluwer Academic Publishers. Printed in the Netherlands.*

After a decade of private practice, Percival was appointed physician to the Manchester Infirmary. As hospitals became more exclusively concerned with the care of the sick during the eighteenth century, they provided a new arena for the struggles of British physicians, surgeons, and apothecaries. After verbal and emotional conflicts persisted among the physicians and surgeons of the Manchester hospital, Percival was asked – in 1792 – to draft a code of rules to regulate and govern practitioners at that hospital.

Since he had no vested interest in the principal London guilds, Percival could view the nexus of traditional relationships as an outsider, so to speak. He realized that the moral statutes of the various colleges of practitioners exerted little significant influence in hospital practice. But, by altering and expanding these statutes, particularly those of the Royal College of Physicians of London, and by using the highest moral sentiments of the age in dealing with the practical problems of hospital practice, Percival cleverly adapted guild regulations to the hospital setting. These rules were accepted by the trustees two years later, and they eventually became the first chapter of a book on medical ethics. After adding three chapters – one about private practice, one about relationships with apothecaries, and one about the legal duties of practitioners, Percival published his *Medical Ethics* in 1803 [12].

There were major differences between the ideals of Gregory and the precepts of Percival. In view of Gregory's thorough discussion, Percival probably thought it superfluous to devote much attention to educational and scientific ideals. Besides, Percival did not believe that an "academical" education was absolutely necessary for a medical practitioner even though he himself was a scholar and an ardent proponent of the experimental philosophy.

Gregory had offered a few standards about interactions of practitioners and about patient care. In appreciating the centrality of consultation in both private and hospital practice, Percival offered many precepts about transactions between physicians, surgeons, and apothecaries. In contrast to the guild statutes, though, Percival emphasized that all matters of consultative decorum should be judged in terms of better patient care. For example, in rural areas, apothecaries usually knew considerably more about the patient than anyone else. Thus, consultation and cooperation between physician, surgeon, and apothecary were desirable not only for professional improvement, but also for a more judicious decision about the care of the patient. In fact, ideals about the conduct of practitioners towards patients were pre-eminent in Percival's code.

There was a third major difference between Gregory's values and those of Percival. Gregory had urged laymen to devote their attention to basic problems of health and disease, but he had not discussed the social obligations of medical practitioners. Percival not only attended to public health obligations, but he also recognized the ethical significance of legal requirements. In return for exemptions from military service and jury duty, for example, physicians were required to meet certain demands of society, including testimony at judicial proceedings requiring medical evidence. This recognition of the impact of laws on professional ethics was a unique contribution by Percival. When applied to medicine, "jurisprudence" meant primarily forensic medicine to the majority of British physicians who practiced in the eighteenth and nineteenth centuries. "Jurisprudence" signified moral injunctions to a few, including Percival and Michael Ryan.

A medical graduate of Edinburgh and a member of the College of Physicians in Edinburgh and in London, Ryan was editor of the *London Medical and Surgical Journal* in the 1830s. In 1831, he published a manual of medical jurisprudence [13], and, five years later, he issued an enlarged edition. The three sections of Ryan's compendium dealt respectively with medical ethics, laws in Britain relating to medicine, and forensic medicine.

Written primarily as a text for students, Ryan wished to prepare a "concise and comprehensive compendium of the moral and legal duties of a medical man." In the section on moral duties, Ryan essayed a history of medical ethics – probably the first in the English language but he did not analyze basic problems of professional values, as had Gregory and Percival. Nevertheless, Ryan is singularly significant because he attempted to correlate medical ethics, health legislation, and forensic medicine. He realized that any society could incorporate its values about professional behavior into civil statutes and, consequently, impose both moral and legal obligations on professional persons. Moreover, practitioners could not satisfactorily discharge professional obligations without understanding community expectations embodied in laws, and a satisfactory fulfillment of certain community obligations involved a special knowledge of law as well as medicine. Thus, Ryan sustained Percival's emphasis on the moral import of laws as well as his understanding of the forensic responsibilities of practitioners.

Various editions of Gregory's monograph, Percival's code, and Ryan's manual appeared between 1770 and 1850. In using these, British and American physicians began to deal with problems of medical ethics in a more thoughtful and organized fashion.

Physicians in the United States who considered problems of medical ethics were well acquainted with the writings of these three men. Benjamin Rush had read Percival's *Medical Ethics*, but, above all else, Rush had been profoundly influenced by his teacher, Gregory. As a professor in Philadelphia, Rush gave at least seven lectures about particular aspects of medical ethics. For example, in 1789 and again in 1801, Rush lectured about the immorality of scientific falsehoods in medicine. Medicine would be vastly improved if the scientific causes that retarded medical progress were removed. Rush repeated some of the causes that Gregory had listed in 1772, and he added some of his own [14].

After Gregory's lectures were reprinted at Philadelphia in 1817, his influence became even greater. Hugh Hodge, in an oration to the Philadelphia Medical Society, reiterated Gregory's suggestions about the importance of certain subjects in medical education. Hodge had also studied Percival's book, and an abridgment of the same was published at Philadelphia in 1823. In that same year, the New York State Medical Society adopted its first code of medical ethics [11]. However, it was not the first American code.

A few rules of professional ethics had been included in the by-laws adopted by some of the medical societies established in the United States before 1800. During the first half of the nineteenth century, these norms were frequently separated from the main group of by-laws and incorporated into codes of ethics, etiquette, or police. The first code was adopted by the Boston Medical Association in March of 1808. Known as the *Boston Medical Police* [15], this code had been prepared by a committee of doctors who claimed that they used the writings of Gregory, Percival, and Rush. Actually, all of the precepts in the *Boston Medical Police* could be found in the second chapter of Percival's *Medical Ethics*, the chapter that discussed such situations in private practice as consultations, arbitration of differences, interferences with another's practice, fees, and seniority among practitioners. Furthermore, the Boston physicians did not explain their reasons for ignoring Percival's precepts about hospital practice, apothecaries, and laws. Although there were few apothecary-practitioners in the United States at this time, there were hospitals, druggists, and medically-related laws. In spite of these exclusions, or, perhaps because of them, the *Boston Medical Police* became *the* model for codes of medical ethics adopted between 1817 and 1842 by at least thirteen societies in eleven states, New York not included.

The New York physicians did not simply imitate the Boston code. They included the forensic obligations so important to Percival. On the other hand, they championed Gregory's ideal about the social arrangement of medical practitioners. Percival had desired that a rigid distinction be maintained between physicians, surgeons, and apothecaries whereas Gregory had challenged British practitioners to learn and practices all branches of medicine. In agreeing with Gregory, the New York physicians expressed the sentiments of other American practitioners who saw no value in supporting the British hierarchy of social distinctions.

The code of the New York State Medical Society exerted obvious influence on those who prepared a code for the Medico-Chirurgical Society of Baltimore in 1832 [10]. The committee of Baltimore physicians also used the code of the Connecticut Medical Society (based on the *Boston Medical Police*) and the writings of Percival, Gregory, Rush and Ryan. In that same year, R. E. Griffith of Philadelphia, had issued an American edition of Ryan's *Manual of Medical Jurisprudence*. To the fifth chapter of this edition, Griffith added a synopsis of Rush's list of duties for patients. This synopsis became part of the fifth section of the code adopted in Baltimore and part of the first chapter of the code adopted by the American Medical Association fifteen years later.

Rush had offered his ideals about the obligation of patients in another lecture to students in 1808. According to Rush, patients should select only those physicians who have received a regular medical education. Moreover, patients should select only those doctors who have regular habits of life and are not devoted to company or pleasure at the theater, turf, or chase. Patients should send for the doctor in the morning but be ready to receive him at any time of the day. They should communicate the history of their complaints fully but not relate the tedious or unimportant details. They should promptly obey the doctor's prescriptions. They should express appropriate gratitude and pay their fees promptly. Thus, the Baltimore doctors added a new dimension to American medical ethics by codifying the ideas of Rush regarding the responsibilities of patients toward their physicians.

With the momentum generated by local and state societies and their codes, it is not surprising that one of the earliest resolutions passed by the delegates to the first national medical convention in the United States involved the creation of a code of medical ethics. The committee who drafted this code reported that "a great number of codes of ethics" adopted by different societies in the United States were "all based on that by Dr. Percival" [6]. In preparing their code, the committee attempted to preserve Percival's words, although a "few of the sections were in the words of the late Dr. Rush," and "one or two sentences" were from other writers. On the afternoon of 6 May 1847, the committee's report was adopted as the first *Code of Medical Ethics* for the American Medical Association [1].

The *Code* was divided into three chapters. The first dealt with the duties of physicians to their patients and, vice-versa, the duties of patients to their physicians. The latter section was exclusively Benjamin Rush. The former included summaries of the comments about patient care scattered throughout Percival. Chapter Two reviewed the obligations that physicians had toward each other. It included all of the aforementioned precepts about consultations, interferences, and disagreements. Chapter Three included the obligations of the profession to the public and, conversely, the obligations of the public to the profession. The latter were generalizations based on the requisites of Rush. The former preserved the concerns of Percival and Ryan by obligating American doctors to attend to matters of public health and forensic medicine.

Thomas Percival's grandson, James Haywood, visiting Philadelphia in December of 1848, expressed his appreciation to one of the members of the AMA committee for their use of Percival's moral precepts. "In England," said Haywood, "I believe that my grandfather's *Medical Ethics* are generally looked upon as a standard work on that subject, and it is gratifying that you have honored him with a similar confidence on this side of the Atlantic" [5]. Between 1832 and 1847, the writings, not only of Percival, but also of Gregory and Ryan, had become familiar to American physicians. In 1834, sections from Percival's book had been published in the *United States Medical and Surgical Journal*. In 1836, Gregory's lectures and Percival's book were recommended to candidates for licensing examination by the Massachusetts Medical Society.

Between 1765 and 1847, therefore, interested Americans studied and unquestionably utilized the British heritage bequeathed by Gregory, Percival, and Ryan. This heritage was essential to the beginnings of American medical ethics. In fashioning their ideals, American physicians borrowed many, but not all of the values offered by the British doctors. Furthermore, there was a major paradox in the transmission of professional values from Great Britain to the United States. Americans had rejected the British arrangements of medical practitioners and had grouped together in local, state, and national societies. In preparing codes of ethics for these societies, though, the Americans adopted many ideals about professional conduct that Gregory and Percival had offered to improve relationships between members of the British guilds. Nevertheless, with the adoption of the national code in 1847, American values and ideals returned to influence British doctors.

Influences from the United States became visible as early as the 1830s. When Michael Ryan revised his book on medical jurisprudence in 1836, he transformed Chapter 5 into a section entitled "American Medical Ethics." This chapter was actually a reprint of the talk that had been given to the Philadelphia Medical Society in 1826 by John Godman. Ryan had not prepared a history of "American Medical Ethics," nor had he mentioned the synopsis of Rush's essay on the duties of patients that Griffith had added to the Philadelphia edition of Ryan's book, but Ryan had acknowledged the existence of an "American" medical ethics.

The momentum of American influences was strikingly on the increase in

Great Britain by mid-century. In 1849, a third edition of Percival's *Medical Ethics*, was published in London – inspired by the attention paid to the book on the other side of the Atlantic. In an essay that appeared in the *London Medical Gazette*, W. B. Kesteven quoted several clauses from the AMA *Code* [8]. In the same year, the London publishing firm of John Churchill reprinted the AMA *Code* and some essays on the duties of physicians written by a Boston physician, John Ware. The most important American author, however, was a private practitioner in Connecticut, Worthington Hooker.

In 1849, Hooker published a monograph with the following title: *Physician and the Patient or, a Practical View of the Mutual Duties, Relations, and Interests of the Medical Profession and the Community* [7]. With this book, Hooker became the first American physician to write an extensive interpretation of the AMA code, and the first nineteenth-century American physician to write a comprehensive monograph on the subject of medical ethics.

In 1850, an edition of Hooker's book was published in London. The editor, Edward Bentley, understood the historical significance of Hooker's book. "It has been the subject of common remark," said Bentley in his preface, "that no work upon the mutual duties, relations, and interests of the medical profession and the community has hitherto appeared in England, and considering the many able men capable of performing this task and whose opinions and experiences in such matters would carry weight and add importance to this interesting subject, it certainly is a matter of surprise. . . ." It is difficult to say which was more surprising to Bentley: the fact that no British physician between 1803 and 1850 had written a monograph on medical ethics, or the fact that a Norwich, Connecticut practitioner had written one that so forcefully illustrated the dimension of mutuality or reciprocity in physician-patient relations. Although calling Hooker "William" instead of "Worthington" on the title page of his London edition, Bentley could not change the fact that an American doctor had made a significant contribution to Anglo-American medical ethics.

Hooker believed that physicians had profound obligations to develop the highest of professional skills, especially those involved in clinical observation and evaluation. He also expected the public to correct their errors about professional skills and to learn how to distinguish between good and bad practices. Furthermore, physicians must understand and adhere to all of the rules of professional decorum, and the public must also understand these rules and appreciate the consequences of interfering with the activities of competent practitioners. With these and other analyses of the mutual obligations of physician and patient – of the medical profession and the community – Hooker championed the ingenuity of the AMA *Code* and brilliantly depicted a feature of professional ethics that was not emphasized by Gregory, Percival, or Ryan.

The American novelty was widespread codification culminating in the AMA *Code* of 1847. This code was voluntarily adopted by many state societies during the ensuing eight years. In 1855, the AMA resolved that all state and local societies had to adopt the code if they wished to send delegates to its annual convention. But compulsory acceptance did not guarantee higher standards or uniform enforcement. In 1857, one critic of codification observed that professional conflicts and abuses were as evident in England where there were no codes as they were in America with codes. Hooker might have retorted that the goods of professional life were more recognizable in the United States with codes than in Great Britain without codes. No American claimed that codes guaranteed medical righteousness. Codes simply provided physicians with some knowledge of the difference between right and wrong professional conduct. Without some ideals and some means of institutionalizing them, there would be little chance to alter professional evils anywhere.

Spurred by the AMA code, the British Medical Association attempted to develop a code of ethics. Prior to 1858, at least two committees faltered. At the Edinburgh meeting in July of 1858, a thirty-four-member committee was established with Charles Hastings as chairman and T. Herbert Barker and Alexander Henry as secretary. At the meeting in 1859, Barker was granted additional time to prepare his report. It had not materialized by 1865 ([2], 1858, 1859).

The American efforts afforded British doctors a mirror by which they could judge the relevant and less relevant parts of their own professional values. Perhaps the British practitioners understood the problems of enforcement and compromised professional freedom inherent in codes. Whatever the reasons, Great Britain did not have a nationally accepted set of ethical guidelines by 1865.

In summary, the international exchange of professional ideals between 1785 and 1865 was not exclusively from Great Britain to the United States. Primarily one-way before 1830, the influences began to shift afterwards. By the middle of the nineteenth-century, British practitioners were well aware of developments in the United States, including the adoption of the AMA *Code* in 1847 and the publication of Worthington Hooker's *Physician and Patient* in 1849. After 1850, practitioners in both countries recognized the challenge of a principle of mutual obligations between practitioners and patients and they reciprocally influenced each other as they created and changed their professional values.

BIBLIOGRAPHY

1. American Medical Association: 1847, *Proceedings of the National Medical Conventions, Held in New York, May, 1846, and in Philadelphia, May, 1847*, Philadelphia; this volume, pp. 65–88.
2. *British Medical Journal* 1858, 657–8; 1859, 631.
3. Forbes, R.: 1954, "Medical Ethics in Great Britain," *World Medical Journal* 1, 297–9.
4. Gregory, J.: 1772, *Lectures on the Duties and Offices of a Physician and on the Method of Prosecuting Enquries in Philosophy*, W. Straham and T. Cadell, London.
5. Haywood, J.: 3 December 1848, "Letter to Isaac Hays," located in Isaac Hays's Papers, Library of the American Philosophical Society, Philadelphia, Pennsylvania.
6. Hays, I.: 1847, "Note," *Code of Ethic*, this volume, pp. 73–74.
7. Hooker, W.: 1849, *Physician and Patient; or, a Practical View of the Mutual Duties, Relations and Interests of the Medical Profession and the Community*, Baker and Scribner, New York.
8. Kesteven, W. B.: 1849, "Thoughts on medical ethics," *London Medical Gazette* 9, 408–14.
9. King, L.: 1958, *The Medical World of the Eighteenth Century*, University of Chicago Press, Chicago.
10. Medico-Chirurgical Society of Baltimore: 1832, *The System of Medical Ethics Adopted by the Society, Being the Report of the Committee on Ethics*, Baltimore.
11. New York State Medical Society: 1823, *A System of Medical Ethics, Published by the Order of the State Medical Society of New York*, New York.
12. Percival, T.: 1803, *Medical Ethics; Or, A Code of Institutes and Precepts, Adapted to the Professional Conduct of Physicians and Surgeons*, J. Johnson, London.
13. Ryan, M.: 1831, *A Manual of Jurisprudence, compiled from the best medical and legal works: comprising an account of: The Ethics of the Medical Profession, II. The Charter and Statutes Relating to the Faculty; and III. All Medico-legal Questions, with the latest discussions. Being an Analysis of a Course of Lectures on Forensic Medicine Annually Delivered in London and intended as a compendium for the use of barristers, soliciters, magistrates, coroners, and medical practitioners*, Renshaw and Rush, London.
14. Rush, B.: 1811, in *Sixteen Introductory Lectures*, Philadelphia, pp. 141–65.
15. Warren, J., Hayward, L., and Fleet, J.: 1808, *The Boston Medical Police*, Association of Boston Physicians, Boston, this volume, pp. 41–46.

CHAPTER 7

PETER BARTRIP

AN INTRODUCTION TO JUKES STYRAP'S *A CODE OF MEDICAL ETHICS* (1878)

Jukes Styrap (de Styrap as he was sometimes known) compiled the only important code of medical ethics to be published in Victorian England. Following its initial publication, in 1878, revised and enlarged editions appeared in 1886, 1890, and 1895. These incorporated fresh material, including sections on consultation with homeopaths, railway medical etiquette, medical detectivism, and an appendix on the issue of bulletins from the sickrooms of distinguished patients. The 2nd edition amounted to a major revision for it ran to 56 pages, exclusive of introduction and preface, whereas its predecessor (reproduced here) had filled only 27. Subsequent editions showed more modest changes.

In 1882 Styrap offered his *Code* to the British Medical Association, of which he had been a member since 1856, in the hope that it would gain acceptance as the profession's ethical standard. He anticipated that as such, it would be sent to every newly-elected BMA member. His hopes were dashed, however, when the Association's governing body, the Committee of Council, for reasons which were not disclosed, declined the offer ([2], II, 1882, p. 192). But while the *Code* never had any official standing within the Association, it exercised considerable influence; hence, in 1896 a *British Medical Journal* leader described it as "the usually accepted authority on ethics in the BMA" ([2], II, 1896, p. 401).

In anticipation of complaints that the profession had no need of a written code, the principles of correct conduct being well understood, the preface to Styrap's first edition provided four justifications for his work. First, the regular requests for guidance and advice which appeared in the medical press (Styrap later claimed that the *Code* was intended to assist the young practitioner). Sec-

R. Baker (ed.), The Codification of Medical Morality, 145–148.
© 1995 *Kluwer Academic Publishers. Printed in the Netherlands.*

ond, the example of "our eminently practical American brethren" who had possessed a written code since 1847. Third, the demonstrated interest of the BMA which, in the 1840s and 1850s, had appointed abortive committees, on one of which Styrap himself had sat, to prepare a code of medical ethics "capable of being adopted by the Association"([2], 1858, pp. 657–8; 1859, p. 631). Fourth, the existence of local medico-ethical societies which possessed codes lacking the requisite detail.

Before publishing, Styrap distributed drafts of the *Code* throughout the profession, including to some of its most eminent members. Many of their comments were then incorporated into the final text. The result was, in the words of the *British Medical Journal*'s reviewer, a "very complete code of medical ethics [which] deals in a very comprehensive and, indeed, almost an exhaustive manner...with the principles which should guide medical men." Practitioners "of all grades," he believed, would find the *Code* "a valuable possession" ([2], II, 1878, p. 105). The second edition gained an even more enthusiastic response, the *BMJ*'s reviewer referring to "this excellent little work" written in "stately and old-fashioned...diction" which should be "treasured for its own sake" ([2], I, 1886, p. 213). Oddly, when, at the end of the century, the BMA again decided that it should have a written ethical code – a decision which once more led to nothing – it ignored Styrap's long-established volume.

Of Styrap himself we know little. The *British Medical Journal* published only a brief obituary while the *Lancet*, which never reviewed any edition of the *Code*, failed to notice his death at all. Elsewhere the silence was equally profound. Born on 30 September 1815, Styrap entered Shrewsbury School in 1826, leaving in 1829, after which he was privately educated at Stourport in Worcestershire. The Shrewsbury School register lists him as plain "Jukes Stirrop", which suggests that the names "Styrap" and "de Styrap" may have owed more to snobbery than to ancestry ([1], p. 63). Styrap went on to study medicine at King's College London where he was taught by Sir Thomas Watson (1792–1882) who, it has been said, was "the acknowledged head of the medical profession" in mid-nineteenth century Britain ([3], pp. 291–3). Notwithstanding Watson's earlier death, Styrap dedicated the second and subsequent editions of the *Code* to his former teacher.

Styrap qualified MRCS and LSA in 1839. What became of him over the next few years is uncertain, though it is clear that he spent time in Ireland where he obtained the licentiate of the Royal College of Physicians of Ireland in 1850 (MRCPI, 1879). In the 1850s he set up practice in Shrewsbury, where he remained for the rest of his life. During the same decade he helped found the Salopian Medico-Ethical Society, of which he was secretary. When this society merged with the Shropshire Branch of the BMA, which merger he helped to negotiate, Styrap became honorary secretary of the associated societies ([2], 1859, pp. 278, 395). Appointed physician to the Salop Infirmary in 1859 Styrap held several other hospital posts including as consulting physician to the South Shropshire, Bridgnorth and Montgomeryshire Infirmaries ([2], I, 1899, pp. 1130–1131).

In 1864 Styrap suffered "a severe illness" from the effects of which he never completely recovered. He retired from practice in the following year and converted his Salop Infirmary appointment to a consultancy in 1867. From this it is clear that the *Code*, which appears to have circulated within the Shropshire Medico-Ethical Society for some years before it was published, was compiled as a retirement activity. Certainly, Styrap was for "many years before his death...practically confined" to his house on College Hill, Shrewsbury. Apart from the *Code* he dabbled with other non-clinical medical writings, namely: his book, *The Young Practitioner* (London: H. K. Lewis, 1890), and the pamphlets, *A Tariff of Medical Fees* (1870) and *Medico-Chirurgical Tariffs* (1874); the second of these ran to five editions, the last appearing in 1890. Another retirement activity was the design of a "urinary cabinet," containing test tubes, thermometer, forceps plus other instruments and equipment, an example of which Styrap exhibited at the BMA's annual museum in 1882 ([2], II, 1882, p. 760). In 1899 he contracted influenza, complicated with broncho-pneumonia, and died on 9 April, aged 83 ([2], I, 1899, pp. 1130–1131).

Styrap never claimed that his *Code* was entirely original. His preface acknowledged a debt to the framers of the laws of the Manchester and Salopian Medical Ethical Societies and to other writers, including the Committee of the American Medical Association, of whose compilation he had "largely availed" himself. Above all, he recognized the towering presence of Thomas Percival, noting that all existing ethical codes, American and English, appeared to be based on Percival's ([4]). At first glance Styrap's *Code* may appear to be little more than a re-hash of previous work. Closer examination reveals that it contains many subtle differences with the second, enlarged, edition showing more obvious changes. Broadly, Styrap tended to be more "hardline," less inclined to compromise and tolerance of ethical misdemeanor, than Percival. It is in these differences that the *Code*'s prime importance lies, for they reveal how Styrap updated Percival to meet the conditions of medical practice in mid- and late-Victorian Britain. They are what make *A Code of Medical Ethics* an important and unjustly neglected work.

BIBLIOGRAPHY

1. Auden, J. E. (ed.): 1909, *Shrewsbury School Register*, Woodall, Minshall, Thomas, Oswestry.
2. *British Medical Journal*.
3. Munk, W. (ed.): 1878, *The Roll of the Royal College of Physicians of London*, Royal College of Physicians, London.
4. Percival, T.: 1803, *Medical Ethics*, Johnson & Bickerstaff, Manchester.

JUKES STYRAP

A CODE OF MEDICAL ETHICS

The duties of medical practitioners to the public and to the profession at large, to each other, and to themselves

CHAPTER I. ON THE DUTIES OF MEDICAL PRACTITIONERS TO THEIR PATIENTS, AND THE OBLIGATIONS OF PATIENTS TO THEIR MEDICAL ADVISERS

Section I. – the Duties of Practitioners to Their Patients

Special Rules, Etc. –

1. A medical practitioner should not only be ever ready to obey the calls of the sick, but his mind should be imbued also with the greatness and responsibility of his mission; and his obligations are the more deep and enduring, as there is no tribunal other than his own conscience to adjudge penalties for carelessness or neglect. A "doctor", therefore, should minister to the sick with a due impression of the importance of his vocation: reflecting, moreover, that the comfort, the health, and the lives of those committed to his charge depend, humanly speaking, on his skill, attention, and fidelity. In his deportment, also, he should study so to unite *tenderness* with *firmness* and *urbanity* with *authority*, as to inspire the minds of his patients with gratitude, confidence, and respect.
2. Every case (rich and poor alike) entrusted to the care of a practitioner should be treated with attention, kindness, and humanity. Reasonable indulgence should also be accorded to the mental weaknesses and caprices of the sick. Delicacy must in all cases be strictly observed, and secrecy also, under all but very exceptional circumstances – as, for instance, in a case of threatening insanity, or of pertinacious concealment of pregnancy after seduction,

R. Baker (ed.), The Codification of Medical Morality, 149–171.
© 1995 *Kluwer Academic Publishers. Printed in the Netherlands.*

in which it would probably be the practitioner's duty to communicate his fears to a near relative of the patient; and the familiar and confidential intercourse to which a "doctor" is admitted in his professional visits, should be used with discretion, and with the most scrupulous regard to fidelity and honour. The obligation of secrecy extends beyond the period of professional services; – none of the privacies of personal and domestic life, no infirmity of disposition, or defect of character, observed during professional attendance, should every be disclosed by the medical adviser, unless imperatively required. The force and necessity of this obligation are indeed so great, that professional men have, under certain circumstances, been protected in their observance of secrecy by courts of justice.

3. In many cases, frequent visits to the sick are necessary, as they enable the medical attendant to arrive at a more perfect knowledge of the disease, and to meet promptly any change of symptoms: they may also, in some instances, be requisite to inspire the patient with confidence; but unnecessary visits are calculated to diminish the authority of the practitioner, and render him liable to be suspected of interested motives, and thus discredit the profession.
4. A practitioner should not be prone to make gloomy prognostications, inasmuch as, they not only exert a depressive influence on the invalid, but savour strongly of empiricism by unduly magnifying the importance of his services in the treatment or cure of the disease; at the same time, he should not fail to give to the friends of the patient timely notice of actual danger, and even to the patient himself, if absolutely necessary, or when specially desired by the relatives. The communication, however, when personally made by the doctor, is generally so alarming to the patient, that, whenever it can, it had better be delegated to some discreet relative, or other sympathising friend; for the medical attendant should be the minister of hope and comfort to the sick – that, by such cordials to the drooping spirit he may soothe the bed of death, revive expiring life, and counter-act the depressing influence of those maladies which often disturb the tranquillity, even of the most resigned, in the trying moments of impending dissolution. Nor should it be forgotten that the ebbing life of a patient may be shortened not only by the acts, but also by the words and manner of the doctor; it is, therefore, his duty carefully to guard himself in this respect, and to avoid, as far as possible, everything which has a tendency to discourage the patient and depress his spirits.
5. A practitioner is not justified in abandoning a patient because the case is deemed incurable; for, even in the last stage of a fatal malady, his continued

attendance may prove highly beneficial to the patient, and a comfort to the sorrowing relatives, by professional suggestions for the alleviation of pain, and the soothing of mental anguish and distress. And here it may be well to note that, but few practitioners, – if any, indeed, save those who have themselves languished on a bed of sickness or, it may be, of apprehended death, – can fully realise the feeling of comfort and consolation afforded by the presence of a kind, sympathising doctor in the chamber of the sick and the dying. To decline attendance under such circumstances, would be sacrificing to ideal delicacy and mistaken liberality, that moral duty, which is independent of, and far superior to all pecuniary consideration. At the same time there are circumstances which fully justify a medical man relinquishing the care of a patient – such as willful, persistent disregard of his advice; the abuse of his attendance as a "blind" for some unworthy purpose, or irregularity of life; loss of the necessary professional restraining influence; and other positions which the practitioner's innate feeling of self-respect will at once indicate, should the necessity arise.

6. In difficult or protracted cases, consultations should be freely and judiciously promoted, as they engender confidence, evoke energy and give rise to more enlarged views in practice.
7. The opportunities which a medical man not infrequently enjoys of promoting and strengthening the good resolutions of a patient suffering from the consequences of alcoholism, or vicious conduct, should never be neglected. His counsels, and even his remonstrances, will generally be taken in good part especially by the younger members of a family, – and give satisfaction rather than offence, if tendered with feeling courtesy.

Section II. – the Obligations of Patients to Their Medical Advisers

1. The members of the faculty, on whom devolve so many important, arduous, and anxious duties on behalf of the community – in the discharge of which, moreover, they have continually, in the interest of the sick, to sacrifice their rest, comfort, and health, and expose themselves to the risks of fevers, and other infectious diseases, – are justly entitled to expect from, and, if need be, should impress upon their patients a due sense of their moral (irrespective of all pecuniary) obligations to the faculty: for it cannot be doubted that the medical profession, characterised as it is by unselfish devotion of life to the necessities of an exacting, and, too often, selfish public, is worthy of the honour accorded to it in the Apocryphal writings: – "Honour a physician with the honour due unto him for the uses which ye

may have of him: for the Lord hath created him. – For of the most High cometh healing, and he shall receive honour of the King. – The skill of the physician shall lift up his head: and in the sight of great men he shall be in admiration," etc. – Ecclus., ch. xxxviii.

2. The first professional (so to speak) duty of a patient is to select, as his medical adviser, a duly educated and registered practitioner. In no profession, trade, or occupation do mankind rely on the skill of an untaught artist; and in medicine, confessedly the most difficult and intricate of the sciences, the world must not suppose that knowledge is intuitive.
3. A patient will do well to elect a practitioner whose habits of life are regular, and not unduly devoted to company, pleasure or other pursuits incompatible with his professional obligations. He should also, as far as possible, confide the care of himself and family to one practitioner: for a medical man who has acquired a knowledge of their constitution, habits, and predispositions, is more likely to be successful in his treatment than one who lacks it.

   Having thus chosen his doctor, a patient will act wisely in applying for advice in cases which, to him, may appear trivial – for serious, and even fatal results not unfrequently supervene (if neglected) on accidents seemingly slight; and it is of still greater importance that he should seek it in the early stage of acute disease: to neglect of this precept is doubtless due much of the uncertainty and failure with which the medical art has been reproached.
4. Patients should faithfully and unreservedly communicate to their medical adviser the supposed cause of their malady. It is the more important, since many diseases of mental origin simulate those dependent on external causes, and yet are incurable otherwise than by ministering to the mind diseased. A patient, moreover, should never be afraid of thus making the doctor his friend and confident, but should always bear in mind that a medical man is under the strongest ethical obligations of reticence and secrecy; nor should any undue feeling of shame or delicacy deter even females from disclosing to him the seat, symptoms, and suspected causes of any ailment peculiar to their sex; for however commendable and necessary a modest reserve may be in the ordinary occurrences of life, its too strict observance in medicine might be attended with the most serious consequences – and a patient may even sink under a painful and loathsome disease, which might have been cured, or, at least, relieved, and much suffering averted, if timely intimation had been given to the medical attendant.
5. A patient, when narrating the symptoms and progress of his malady, should avoid unnecessary prolixity and detail which would weary the attention

and waste the time of his doctor; neither should he, without good cause, obtrude upon him the details of his business, nor the history of his family concerns. Even as regards his actual symptoms, he will convey much more real information by giving clear answers to interrogatories, than by the most minute self-statement.

6. The obedience of a patient to the prescriptions and instructions of his medical adviser should be prompt and implicit, and his attention to them uninfluenced by his own or other crude opinions, as to their fitness – for a failure in any one particular may render an otherwise judicious plan of treatment hurtful, and even dangerous. Nor can caution be too strongly impressed upon convalescent patients, who are very apt to suppose that the rules prescribed for them may then be disregarded – and the not uncommon result is a relapse, consequent on some indiscretion in diet, exercise, or undue exposure. – Patients, moreover, should never allow themselves to be persuaded to take medicines recommended to them by the self-constituted doctors and doctresses so frequently met with in society, and who assume to possess infallible remedies for the cure of this or that disease. However simple their assumed remedies may seem to be, it not infrequently happens that they are productive of much mischief, and in all cases are likely to be injurious, by contravening the treatment and impairing the authority of the medical attendant.
7. A patient should avoid even the *friendly visits of a practitioner* not in attendance upon him; and if constrained to receive them, *he should never converse on the subject of his malady* – for an observation might be made, which, without any intention to professionally interfere, may weaken or destroy his confidence in the treatment pursued, and induce him to neglect the directions laid down for his guidance.
8. The confidential relations which usually subsist between patient and practitioner render it especially incumbent on the former, during illness, to be open and unreserved with his medical adviser; and he ought never to send for a consultant without the knowledge of his ordinary medical attendant. It is also of great importance that practitioners should act in concert; for although their respective plans of treatment, if carried out singly, may be attended with equal success, yet if conjointly adopted, they are very likely to be productive of disastrous results.
9. Patients should always, when practicable, send for their doctor in the morning, before his usual hour of going out; for by an early knowledge of the visits he has to make during the day, he is enabled so to apportion his time as to obviate any clashing of engagements. They should also avoid calling

on, or sending for him during the hours devoted to meals or to sleep, unless really necessary. They should, likewise, always endeavour to be ready to receive his visits, as detention, even for a few minutes, is often of serious inconvenience to a practitioner in extensive practice: – on the other hand, the medical attendant will do well, even if it be not a duty incumbent upon him to intimate as nearly as may be, the hour at which he intends to make his next visit – for most patients not only like to have their persons, and their rooms tidied for his reception, but the protracted anxious expectancy, and long for the doctor's rap has, there is little doubt, "like hope deferred", a prejudicial effect on the recovery of the sick.

10. Patients should, after their recovery, entertain a just and enduring sense of the value of the services rendered to them by their doctor; for, in severe illnesses especially, these are usually of such an anxious, trying nature, that no mere pecuniary acknowledgment can repay or cancel them.

CHAPTER II. – ON THE DUTIES OF MEDICAL PRACTITIONERS TO THE PROFESSION, TO EACH OTHER, AND TO THEMSELVES

Section I. – the Duties of Practitioners in Support of Professional Character and Status

1. Everyone who enters the profession, and thereby becomes entitled to its privileges and immunities, incurs the obligation to exert his abilities to promote its honour and dignity, to elevate its status, and extend its influence and usefulness. He should, therefore, strictly observe such laws as are instituted for the guidance of its members, and avoid all disparaging remarks relative to the faculty as a body, or its members individually; and should seek by diligent research and careful study to enrich the science and advance the art of medicine.
2. There is no profession, from the members of which greater purity of character, and a higher standard of moral excellence, are required, than the medical; and to attain such eminence is a duty which every practitioner owes alike to his profession, and to his patients. It is due to the latter, in so far, that, without it, he cannot command their confidence and respect: and to both, since no scientific attainments can compensate for the want of sound principles of morality. It is also incumbent upon the faculty to be temperate in all things – for the practice of physic requires the unremitting exercise of an unclouded and vigorous understanding; and on emergencies (for which

no professional man should be unprepared), a steady hand, a quick eye, and a clear head, may be essential for saving the life of a fellow creature.

3. It is degrading to the true science of medicine to practise . . . professedly or exclusively, hydropathy or mesmerism; and alike derogatory to the profession to solicit practice by advertisement, circular, card, or placard; also, to offer, by public announcement, gratuitous advice to the poor, or to promise radical cures; to publish cases and operations in the daily press, or knowingly, to suffer such publications to be made; to advertise medical works in non-medical papers; to invite laymen to be present at operations; to boast of cures and remedies; to adduce testimonials of skill and success; or to do any like acts. Such are the ordinary practices of charlatans and are incompatible with the honour and dignity of the profession.
4. Equally derogatory to professional character is it for a practitioner to hold a patent for any proprietary medicine or surgical instrument; or to dispense a secret *nostrum*, whether it be the composition, or exclusive property of himself, or of other: for, if such nostrum be really efficacious, any concealment in regard to it is inconsistent with true beneficence and professional liberality; and if mystery alone impart value and importance to it, such craft is fraudulent. It is also extremely reprehensible for a practitioner to attest the efficacy of patent or secret medicines, or, in any way, to promote their use; only less culpable is the practice of giving written testimony in favour of articles of commerce, and tacitly or otherwise sanctioning its publication. It is likewise degrading for a medical man to enter into compact with a druggist to prescribe gratuitously or otherwise, and, at the same time, share in the profits arising from the sale of the medicines. Alike censurable (and ethically dishonest) is the modern practice of assuming, for the purely selfish purpose of personal advancement, the distinctive titles and status of our public institutions, and parading private speculations as *bona-fide* "hospitals," "infirmaries," and "dispensaries." Such *sham* institutions are not only derogatory to the faculty, but injurious to the true interests of the community; and no practitioner desirous to uphold the dignity of his profession should resort to such *unprofessional* devices – otherwise he must not be surprised at being ignored by the faculty and treated as a charlatan.

Section II. – the Duties of Practitioners in Regard to Professional Services to Each Other

1. All legitimate practitioners of medicine, their wives, and children while under the paternal care, are entitled to the gratuitous (traveling expenses

excepted) services of any of the neighbouring faculty, whose assistance may be desired.

A doctor suffering from serious disease is, in general, an incompetent judge of his own case: and the natural anxiety and solicitude which he experiences at the sickness of a wife, child, or others, who, by the ties of consanguinity, are rendered dear to him, tend to obscure his judgment, and engender timidity and irresolution in his practice. Under such circumstances, medical men are especially dependent upon each other: and kind offices and professional aid should always be cheerfully and freely afforded. Visits should not, however, be officiously obtruded, since unsolicited attention may give rise to embarrassment, or interfere with that choice on which confidence depends. But if a member of the faculty, in affluent circumstances, request attendance, and an honorarium be tendered, it should not be declined – for no pecuniary obligation ought to be imposed on the debtor, which the debtee himself would not wish to incur.

Offices. Section III. – the Duties of Practitioners in Respect to Vicarious Offices

1. The affairs of life, the pursuit of health, and the various accidents and contingencies to which a medical man is peculiarly exposed, sometimes necessitate a temporary withdrawal from practice, and an appeal to some one or more of his professional brethren to officiate for him. A ready assent to such request, or a cordial tender of service when the necessity for such is known or felt, is an act of Christian duty, which, on the divine principle of "Whatsoever ye would that men should do to you, do ye even so to them," should always (if it be possible) be courteously accorded, and carried out with the utmost consideration for the interest and character of the "medical brother." – But if a practitioner neglect his professional duties in quest of pleasure and amusement, he is neither morally nor ethically entitled to the exercise of such fraternal courtesy without adequate remuneration being made to his officiating friend for the services rendered.

Section IV. – the Duties of Practitioners in Consultations

1. The possession of a Degree or Diploma specified in Schedule A of the Medical Act, 1858, furnishes the only presumptive evidence of professional abilities and acquirements, and ought to be the only acknowledged right of an individual to the exercise and honours of his profession. Nevertheless,

inasmuch as in consultations the good of the patient is, or should be, the sole object in view, and that such often depends on personal confidence – no intelligent qualified practitioner possessing a Degree or Diploma from a Foreign, Colonial, or Indian University, of known (though not officially recognised in Great Britain) reputation, and who is, moreover of good moral and professional local standing, should be fastidiously excluded from fellowship, or his aid refused in consultation when it is particularly desired by the patient. But no one can be considered a regular practitioner, or a fit associate in consultation, whose practice is based on an exclusive dogma, such as homeopathy, *et hoc genus omne*, (unqualified Assistants included): – indeed, for a legitimate or orthodox practitioner to meet a professor of homeopathy in consultation, is a dishonest and a degrading act: – dishonest, because he lends his countenance to that which he knows to be a dangerous fallacy – and degrading, inasmuch as he has neither the manly, professional honesty to resist the temptation of a possibly liberal fee, nor the moral courage to discountenance the capricious vagaries of some wealthy, or, may-be, titled patient.

2. It cannot be too strongly impressed on every member of the profession, that in consultations, all feelings of emulation and jealousy should be carefully laid aside; that the most honourable and scrupulous respect for the character and standing of the practitioner in charge of the case should be observed; that the treatment of the latter, if necessary, should be justified as far as it can be consistently with a conscientious regard for truth – and no hint or insinuation thrown out which could impair the confidence reposed in him, or otherwise affect his reputation. The Consultant should also carefully abstain from any of those inordinate attentions, which have been sometimes practised by the unscrupulous for the purpose of gaining undue credit, or ingratiating themselves into favour.
3. In consultations,[1] it is the rule and custom for the Consultant, after the usual preliminary conference relative to the history and facts of the case, to take precedence of the family doctor in the necessary physical and questionary examination of the patient: – exceptional circumstances, however, may arise, in which the family attendant, should, as an act of confidence and courtesy, be the first to propose the necessary questions – after which, the Consultant should make such further enquiries and examination as he may deem necessary to satisfy himself of the true nature of the case; but no observations of any kind indicating an opinion as to the nature of the malady, treatment pursued, or its probable issue, should be made in the hearing of the patient, or his friends, until the consultation is concluded. Both practitioners should

then retire to private room for deliberation; and the treatment having been determined by the consultation of himself and colleague, the Consultant last called in (if there be more than one in attendance), should write the prescription for the medicines decided on – with the name of the patient and the date, – and append his initials thereto, and be followed by those of his colleagues in the order in which they attended. He (the consultant) should likewise be the one to communicate to the patient, or his friends, the directions agreed upon, together with any opinion it may have been decided to express: but no statement should be made, or discussion relative thereto take place before the patient or his friends, except in the presence, and with the consent of all the faculty in attendance; and no *opinions* or *prognostications*, other than those mutually assented to after deliberation, should be expressed.

4. In consultations, and in cases where the ordinary family attendant visits the patient more frequently than the Consultant, it will be his duty to see the measures agreed upon faithfully carried out – not to add to, diminish, or alter, in any way, the practice mutually assented to – except in an emergency, or unexpected change in the case; and in such latter event, any variation of the treatment should, with the reasons for it, be fully explained at the next consultation. The same privilege and duty devolve on the Consultant, when sent for in the absence of the regular attendant.
5. When two, or more, practitioners attend in consultation and the hour of meeting has been fixed, punctuality should be strictly observed; and tis, in most instances, is practicable – for society is, in general, so far considerate as to allow the plea of a professional engagement to take precedence of all other. An unlooked for accident, or other urgent case, may, however, intervene, and delay one of the parties; in that case, the first to arrive should wait a reasonable time for his associate – after which, the consultation should be considered as deferred until a new appointment can be made. If the attending practitioner be the family doctor, he will of course see the patient and prescribe; but if it be the Consultant, he should retire, except in a case of urgent necessity, or when he has been summoned from a long distance – under which circumstances, he may examine the patient, and express his opinion *in writing* (if necessary) and *under seal*, to be delivered to his associate, – and, in the interim, should meet the emergency by such treatment as he may deem necessary.
6. When a senior practitioner is called upon to meet his junior in consultation, for a second opinion it will be competent for the former to represent the propriety and advantage of obtaining the assistance of a more experienced

practitioner; but if the patient specially desire to have the opinion of any qualified member of the profession, even though a junior, it will be at the option of the practitioner in attendance to acquiesce, or withdraw. As a rule, however, a practitioner should never decline to meet another, *merely* because he is his junior; and he will best consult his own interest and that of the profession, by a ready and courteous assent to meet any junior of good repute: – a contrary course would reflect discredit on himself and the faculty.

7. In consultation, the graduate in medicine practising as a physician only, is entitled to precedence of the general practitioner.
8. If, when more than two practitioners have met in consultation, an irreconcilable diversity of opinion unfortunately occur, that of the majority should be acted upon; but if the members on either side be equal, then the decision should rest with the family attendant: in either case, the greatest moderation and forbearance should be observed, and the fact of the disagreement communicated to the patient, or his friends, and the issue left to them. It may also happen that, in the ordinary dual consultation, the two practitioners fail to agree in their views of a case, and the treatment to be pursued – an incident always to be much regretted, and, if possible, avoided by such mutual concessions as are consistent with the dictates of judgment. If, nevertheless, a difference of opinion exist, it would be well to call in a third practitioner: and if that be impracticable, it must be left to the patient to select the one in whom he would wish to confide. At the same time, as every practitioner justly relies upon the rightness of his judgment, he should, when unable to concur in the treatment adopted, consistently and courteously retire from any further participation in the consultation, or management of the case, unless exceptional circumstances should, in the interest of the patient, render such a course undesirable.
9. In consultations, theoretical disquisitions should be studiously avoided, as they often lead to perplexity and loss of time. Consultative discussions, moreover, should be regarded as private and confidential: and neither by word nor manner should any of the parties to a consultation covertly allege, or in any way intimate to the patient, his friends, or other person, that he had dissented from the treatment as unsuited to the case. A proceeding so unethical would not only be dishonouring to the individual practitioner, but a reflection on the faculty. The responsibility, and imputation of failure, however unjust, should, equally with the credit of success, be shared alike by the respective practitioners.

10. Whenever a "second opinion" is desired or suggested by a patient, or his relatives, it should, as a rule, be at once courteously acceded to by the attending practitioner – who too often demurs, or unwillingly assents, under the *erroneous* impression that a consultation detracts from his professional status, and evinces personal distrust in himself: – whereas, it should be regarded simply as the very natural desire on the part of the relatives to leave nothing undone that might perchance, however forlorn the hope, tend to restore the health, or, it may be, save the life of the loved one – cost what it may. But even were it otherwise, it must not be forgotten that the patient has an indisputable right to "further advice", if he wishes it; and the family attendant will do well for is own sake, as well as that of the patient, to let the responsibility be shared by a second practitioner.
11. When from any cause the continued attendance of two practitioners would be objectionable to the patient, and a special and exhaustive consultation – entailing an unusual sacrifice of time – is, in consequence, deemed desirable, a double fee may fairly be charged; and in difficult and obscure cases, and complicated railway and other injuries, in which a minute physical or other examination and a prolonged consultation are rendered necessary, it is only reasonable that the honorarium should be proportionate to the time occupied – as is customary with "Counsel:" an exclusive fee, therefore, of from two to live guineas, according to the social and pecuniary position of the patient and the professional status of the Consultant, may be justly claimed. Due intimation, however, of the Consultant's expected fee in such cases should be given to the patient by the family attendant, prior to the consultation being arranged.
12. The Consultant has no claim to be regarded as a regular attendant on the patient; and his attendance ceases after each consultation, unless otherwise arranged. The patient and his ordinary medical adviser are therefore fully at liberty to call in any other Consultant without the cognizance of the former, provided that no appointment then exists.

    N.B. – Should the practitioner who has been called in consultation be subsequently requested to take sole charge of the patient, he should courteously but *firmly* decline.
13. No member of a firm of practitioners (unless, from professional status and experience, his ordinary personal practice has become purely "consultant," *and his advice, as such, be* specially *requested by the patient*), whose opinion is sought in a case under the care of a partner in the firm, is entitled, according to professional usage, to claim the customary fee of a

Consultant: – such advisory visits, indeed, (if within the prescribed distance of an ordinary visit,) are generally regarded as complimental ones.

Section V. – the Duties of Practitioners in Reference to Substitutes or Locum-Tenentes, and Incidental Interference with Other Than Their Patients

1. Medicine as an art and science is a liberal profession, and those admitted into its ranks should found their expectations of success in practice on the nature and extent of their scientific and personal qualifications, and not on artifice or intrigue.
2. When a practitioner from motives of friendship, or the necessities of business, is prompted to visit a patient under the professional care of another, he should observe the strictest caution, circumspection, and reserve. No meddling enquiries should be made, no disingenuous hints given relative to the nature and treatment of the disease, nor any line of conduct pursued that may directly or indirectly tend to diminish the confidence reposed in the family attendant. indeed, such visits should be avoided, except under peculiar circumstances; and, when made, the topics of conversation should be as foreign to the case as possible.
3. When during sickness, affliction, or absence from home, a practitioner entrusts the care of his practice to a professional friend, the latter should not make any charge to the former, or to the patients for his services, but should in all things be the locum tenens of the absentee. If, however, the attendance be protracted, and the labour proportionate, a fitting acknowledgment should, if circumstances admit, be made.
4. When a practitioner attends for, or in consultation with another, and it appears necessary to change the treatment, it should be done with the most scrupulous care, so as not to injure the reputation or wound the feelings of the previous attendant. Unnecessary, meddlesome interference with the treatment should be carefully avoided as unjust to the family doctor, and derogatory to true science.[2]
5. When a practitioner is consulted by a patient whom he has previously attended as the officiating friend of another during sickness or absence from home, he should act in strict accord with the principle laid down in Rule 9, and decline attendance, except in consultation.
6. When a practitioner is ill or absent from home, and the patient wishes to have a medical man of his own choice, rather than the officiating friend, the practitioner so elected should act in accordance with the following rule: –
7. When a practitioner is called to an urgent case in a family usually attended

by another, he should (unless his assistance in consultation be desired,) when the emergency is provided for or on the arrival of the attendant in ordinary, resign the case to the latter – but he is entitled to charge the family for his services.

8. Whenever a patient, whose usual medical adviser resides at a distance, sends for a practitioner residing near, the latter should adhere to the preceding rule, as far as circumstances admit.
9. When a practitioner is called in to, or consulted by a patient who has recently been, or still may be, under the care of another for the same illness, he should on no account interfere in the case, – except in an emergency – but request a consultation with the gentleman in previous attendance. If, however, the latter refuse this, or has relinquished the case, or if the patient insist on dispensing with his services, and a communication to that effect be made to him, the practitioner last consulted will be justified in taking charge of the case. Under such circumstances, no unjust or illiberal insinuations should be thrown out in reference to the conduct or practice previously pursued – which, as far as candour and regard for truth and probity will permit, should not only be justified, but, if right, honourably persisted in; for it often happens that, when patients from the treatment, they become dissatisfied, and, under the impression that their case is not understood by the "doctor," unjustly impute the blame to him; many diseases, moreover, are *per se* of so protracted a nature, that the want of success in the early stage of treatment affords no evidence of a lack of skilled professional knowledge.
10. When a practitioner is consulted at his own residence, it is not necessary for him to enquire if the patient is under the care of another. It is better, however, that he *should* make the enquiry, and propose a consultation, or communication with the practitioner (if there by any) under whose care the patient has previously been.
11. When a practitioner is called upon by the assistant, or servant of another, to attend to an accident or other emergency in a family to whom both are equally strangers, the former is not entitled to take charge of the case throughout, but should act and be remunerated in conformity with Rule 7, and resign the case.
12. When a practitioner is called in to attend at an accouchement for another, and completes the delivery, or is detained for a considerable time, he is entitled by custom (except in the case of illness, etc. provided for by Rule 3) to one-half of the fee; but on the completion of the delivery, or on the arrival of the pre-engaged accoucheur, he should resign the further man-

agement of the case. In a case, however, which gives rise to unusual fatigue, anxiety, and responsibility, 'tis right that the accoucheur in attendance should receive the entire fee. Note. – In either event, when the officiating accoucheur is a stranger, or a non-acquaintance of the family doctor, the full fee should be tendered to him.

13. When a practitioner has officiated for, or been called in consultation by another, and the ordinary medical attendant has resumed exclusive attendance upon the case, the former should not under any pretext, make friendly calls upon the patient, unless justified by previous personal intimacy: such visits, even in the latter case, would be better omitted for a time.
14. A practitioner, when on a professional visit in the country, may be requested to see a neighbouring patient who is under the care of another. Should this arise from any sudden change of symptoms, or other pressing emergency, he will be justified in giving advice adapted to the circumstances (the nature, which he should in person or by note, at once communicate to the attending practitioner), but should not interfere further than is absolutely necessary with the general plan of treatment, nor assume a future direction of the case, except in consultation with the family adviser, or by special desire of the friends – in which latter event, he should act in accordance with the principled expressed in Rule 9.
15. In cases of sudden illness, or of accidents and injuries, it frequently happens, owing to the alarm and anxiety of friends, that several practitioners are simultaneously sent for. Under these circumstances, courtesy should assign the patient to the first who arrives, and he should select from those in attendance any additional assistance that may be necessary. In all such cases, however, the officiating practitioner should request that the family doctor (if there be one), be summoned; and, unless his further attendance be desired, should at once resign the case to the latter on his arrival.
16. In a case of sudden or accidental death, in which the deceased person was incidentally attended by a practitioner other than the usual "family doctor" – the latter, in the event of a post mortem examination being deemed necessary, should be specially invited to be present: a contrary course would be highly discourteous and censurable.
17. It sometimes occurs that a medical man has the case of a patient under the care of another practitioner stated to him in so direct a manner, as to render it difficult to decline attention to it. In such an event, his observations should be made with the most delicate propriety and reserve. On no account should he interfere with the curative plans pursued, except in cases

where artful ignorance seeks to impose on credulity, – or where neglect, or rashness, threatens the patient with imminent danger.

18. A wealthy or retired practitioner should abstain from giving gratuitous advice to the affluent or "well-to-do" – for to dispense with fees which may justly be claimed is not only a default of duty to the profession, but, to a certain extent, a defraudment of the faculty by the patient and the practitioner.

    [Note. – By the expression – "patient of another practitioner," – is meant a patient who may have been under the care of another practitioner at the time of the attack of sickness, or departure from home of the latter, or who may have requested his professional attendance during such absence or sickness, or in any other manner given it to be understood that he regarded the said practitioner as his regular medical attendant.]

Section VI. – the Duties of Practitioners When Differences Occur between Them

1. When a diversity of opinion, or opposition of interest, occasions controversy and contention between medical practitioners, the matter in dispute should be referred to the arbitration of one or more physicians, surgeons, or general practitioners, as may be mutually agreed upon, – or to three practitioners – one to be nominated by each disputant, and the third by the selected two, – or, when practicable, to a County "Court Medical"; but neither the subject matter, nor the adjudication, should be communicated to the patient or friends, excepting under special circumstances: – for publicity in cases of ethical disputes (the points involved in which are usually neither understood, nor appreciated by general society) may be personally injurious to the practitioners concerned, and can scarcely fail to bring discredit on the faculty at large.
2. In all cases of arbitration, a written statement of the charges preferred, and a like answer thereto, should be required from the respective disputants – with such affirming or rebutting testimony as may be essential to elucidate the facts of the case; and after giving careful consideration to the evidence adduced, the members of the "Court" should proceed to deliver their opinions in succession, from the junior to the senior, in order that the former may not be unduly influenced by the utterances of the latter.

   As a rule, however, no arbitration should be undertaken until the accusant has, either in person or by note, communicated with the accused on the subject of complaint, and failed to obtain an explanation or redress.

[It may here be well to repeat that experience and observation leave little doubt, that, in numerous instances, professional differences arise from some misrepresentation or suppression of the truth (a fruitful source of the unhappy differences, heartburnings, and jealousies, which too frequently disgrace our profession!) by patients, or their friends, rather than direct unethical conduct on the part of the practitioners. Be that as it may, it is equally the duty of everyone who thinks himself aggrieved to dispassionately consider whether he really is so – for, unhappily, some men are so morbidly sensitive, suspicious and jealous, that even were they to be associated with (so to speak) mundane angels, they would fancy their ground invaded, and their rights and *self* ignored. – A medical man should ever be slow to admit that a brother practitioner has knowingly and intentionally wronged him; a little reflection and reasonableness would often suggest an explanation of conduct that, at first, may seem offensive or selfish. Assuming, however, that he is really injured, – that a neighbouring practitioner has acted unethically, and, mayhap, repeatedly so! What, in such case, is to be done? His duty is certainly, as yet, not to publish to the world his personal quarrel – for professional quarrels are discreditable, and not to be lightly proclaimed. Moreover, when a man is clearly in the right, he can afford to exhaust all gentle means of remonstrance and redress: and, in strict accordance with both scriptural and professional ethics, he should, either in person or by courteous note, "go and tell his brother his fault" privately. Should that fail, and the aggrieved party be ultimately obliged to refer the matter to the arbitration of a mutual professional friend, or to a "Court Medical," even then, his object should be, not that the offender should be "shunned," but effectually rebuked, and convinced of his error. Such object is, in many cases, more likely to be gained by private than by public means. But as there are men in the medical, as in other professions, who can only be effectively influenced by public censure; this, under certain circumstances, would be a perfectly legitimate *dernier ressort* through the action of a "Court Medical."]

Section VII – the Duties of Practitioners in Reference to Professional Changes

1. In the interest of the faculty and of the public, it is desirable that some general authoritative rules relative to professional[3] charges should be adopted in every town or district, for the special guidance of the junior practitioners, who are often in doubt as to the remuneration to which they are fairly entitled. Such rules, 'tis scarcely necessary to remark, should be of a some-

what elastic character (with, at least, a minimum guiding fee), – inasmuch as the charges must necessarily, as a rule, be more or less regulated by local circumstances, the social and pecuniary position of the patients, and, in some degree, by the age and local status of the respective practitioners; and it should moreover, be deemed a point of honour to adhere to such rules with as much uniformity as the varying circumstances will admit.

2. It is alike desirable (bearing in mind that, to the commercial or trade-class of society, quarterly or half-yearly payments are now the rule), to impress upon the faculty the expediency of sending in their usual statement of professional charges *annually* or *bi-annually*: – for the "Doctors," proverbial delay, or neglect in the matter, is often attributed to a wrongful motive, and may, indeed, not unfairly be regarded as an incentive to the feeling so forcibly depicted in the following quaintly truthful lines: –

   God and the Doctor we alike adore
   When on the brink of danger, not before;
   The danger past, both are alike requited:
   God is forgotten, and the Doctor slighted!

   It may also be well briefly to allude to the professionally inherent but injudicious system of *deferred* settlements of account, with its natural sequel – a chronic state of indebtedness of patients – which not infrequently lead to a disruption of friendly feeling, and a loss of practice; nor should it be forgotten, moreover, that many who would willingly pay a semi-annual, or a yearly bill, are oft unable to discharge an accumulated one of two or more years.

3. Should a patient question the accuracy of a "non-itemed" bill, his right to be furnished with a statement as to the number and dates of visits, and the special services charged for, should at once be conceded, and reference to the respective items in the ledger permitted – or, better still, suggested: but the service being acknowledged, no abatement (especially under such circumstances) should be assented to on any plea other than absolute inability to meet it in consequence of poverty, or for a like sufficient reason.

   [That a man should entrust the lives of himself and family to the care of a medical practitioner with entire confidence, and yet deem him capable of making an unjust charge for the anxious and grave responsibility entailed upon him in the discharge of his onerous duty is one of the curious anomalies and inconsistencies existent in the several grades of life, and which it behoves the profession to courteously but firmly resent. Such patients, indeed, are best erased from the practitioner's visiting list.]

CHAPTER III. – ON THE DUTIES OF THE PROFESSION TO THE PUBLIC AND THE OBLIGATIONS OF THE PUBLIC TO THE PROFESSION

Section I. – the Duties of the Profession to the Public

1. It is the duty of the faculty, as good citizens, to be ever vigilant for the welfare of the community, and to bear their part in sustaining its institutions and burdens; they should also be ready to advise the public on subject specially appertaining to their profession – such as public hygiene, legal medicine, and medical police. It is their province to enlighten the public in reference to quarantine regulations; the location, arrangement and dietaries of hospitals, asylums, schools, prisons, and like institutions; also in regard to the medical police of towns, – as drainage, water-supply, ventilation, and sanitation generally; and in respect to measures for the prevention and mitigation of epidemic and contagious diseases; and, when pestilence prevails, it is their duty to face the danger, and to continue their labours for the alleviation of the suffering, even at the risk of their own lives.
2. Medical men should also be ready, when called on by the legally constituted authorities, to enlighten courts of inquisition and justice on matters strictly medical – such as involve questions of sanity, legitimacy, murder by poisons or other violent means, and the various other subjects embraced in the science of Medical Jurisprudence. But in such cases, and especially those in which a critical *post mortem* or other scientific examination is necessary, it is only right and just, in consideration of the time, labour, and skill required, that the responsibility and adequate fee so often tendered, under the plea of legal restriction, should be awarded for the skilled service. [In certain cases, in which the required evidence is not compulsory on the practitioner, it may at times be prudent on his part to stipulate (as is the rule with "Counsel"), for an adequate and specified fee.]
3. In giving evidence on any medical question before a Court of Law, or other tribunal of society – whether in criminal or civil matters, – the faculty should act with thoughtful care and rigid impartiality: –

A. – In "Criminal Cases" – lest their testimony should tend either to prejudice the cause of an innocent person, or lead to a failure of justice.
B. – In "Civil Causes" – as in suits for compensation after railway, or other accident, – that they may not by partial or partisan evidence unintentionally mislead the Court.

With the view to avoid the lamentable differences of opinion which, proclaimed in open court, have undoubtedly brought discredit upon medical evidence in general, and scandal on the profession at large, – it cannot be too forcibly impressed upon the faculty, that, in all such cases, *bona-fide*, honest consultations should be freely held between the professional witnesses of the respective litigants; that differences of opinion should be courteously advanced, and carefully weighed and argued; that each with the other should be frankly ingenuous, and unreservedly open – or in other words, that concealment or mental reservation, in any form, either of facts or opinions, should be scrupulously avoided; and, on the principle that *truth* and *justice* are the sole objects sought by the medical witnesses on either side, all feeling of the advocate or partisan should be thoughtfully eliminated and shunned: – in fine, the skilled witness should never allow his personal feelings to overcome his sense of justice.

4. There is no profession by the members of which eleemosynary services are more liberally dispensed than the medical. Duty to self, however, renders it necessary to impose a limit to such devotement. Poverty, professional brotherhood, and certain of the public duties referred to in the first paragraph of the section, should always be recognised as presenting claims for gratuitous services; but no such privilege can be conceded to Government or State Services, or to institutions endowed by public or private benevolence, or to societies for mutual benefit, or to any profession, guild, or trade, or other "bread-winning" occupation; nor can medical men be expected to furnish certificates of inability to serve on juries, perform militia or other public duties, or to testify to the state of health of patients desirous to insure their lives, obtain pensions, or the like without a fee: *but to individuals in indigent circumstances, such professional services should always be freely and cheerfully accorded.*
5. It is likewise the duty of medical men – who so often become professionally or otherwise cognizant of the malpractices and malversation of charlatans (many of whose victims, from very shame, remain silent on the subject of their sufferings), and of the great injury to health, and loss of life even, caused by the baneful use of quack medicines, – to enlighten the public on the subject, and to judiciously expose the artful devices and unscrupulous pretensions of the charlatanic medical impostor. Practitioners should, moreover, in the interest of the public welfare, exert all their influence to induce chemists, and others, to discountenance the sale and use of empirical or secret remedies, and deter them from being in any way engaged in their manufacture: indeed, so long as they act as the common venders of quack nostrums, and persist in illegal "counter practice," to the detriment of the

public and the faculty, it may safely be affirmed that, as a rule, the body of general practitioners will not have recourse to the chemists as their "dispensers-in-ordinary," but continue the present convenient thought improvable system of "home dispensing."

Section II. – the Obligations of the Public to the Profession

1. The benefits accruing to the public, directly or indirectly, from the active and unwearied beneficence of the profession are so numerous and important, that medical men are justly entitled to the utmost consideration and respect from the community. The public ought likewise to entertain a just appreciation of medical and surgical qualifications; to make a proper distinction between true science and the assumptions of ignorance and empiricism; to afford every encouragement and facility for the acquisition of practical instruction – and not to allow the statute-books to exhibit the anomaly of exacting, purely in the interest of the people, skilled knowledge from the profession under a liability of heavy penalties, and at the same time deter them, by penal enactment, from resorting (except under stringent and unnecessary restrictions – insulting to a proverbially humane profession) to one of the most effective of advancing medical science: – *viz.* carefully conducted experiments on living animals – vivisection, so called – to the scientific practice of which, the present state of medical knowledge, and the advance it has made during the last half century, are, in a great degree, undoubtedly due. If, in absurd deference to the denunciatory and unreasonable clamour of a small, morbid section of society, the scientific use of vivisection were prohibited, the progress not merely of theoretic physiology, but of practical medicine would be greatly retarded, and a serious loss entailed on suffering humanity.

Section III. – the Use of and Property in Prescriptions

1. The common assumption that physician, or other practitioner, in writing a prescription, loses all right of property therein, and that the pharmacist or chemist, who compounds and copies it, acquires a title to use it as he pleases, and the patient the right of perpetually disposing of it, is one so wrong in principle that it demands from the profession greater attention than it has hitherto received. It may, therefore, be well to state that a prescription is neither more nor less than a written order, or direction, to the pharmacist to furnish or compound medicines for the use of the patient, and undoubtedly

remains the property of the author; and neither the patient, nor the pharmacist has any right to use it, excepting for the case and purpose specified: – for although, as Professor Ordronaux has justly remarked, "the party paying for the prescription has an indisputable right to the *personal use* of the formula, he acquires thereby no absolute property in it. That he may use it personally, as often as he pleases, cannot be doubted – for the use is precisely what he purchased: *but he has no right to give it to others*." The respective rights in a prescription, therefore, may be thus briefly defined: – that the physician, as the author, has a literary property in the composition of the formula, and the right to dispose of the use of it to a patient without invalidating his title to the original ownership; that the pharmacist by compounding the same acquires no claim whatever thereto, other than as a record, or justification for dispensing it – in fine, his right is simply that of a custodian; whilst that of the patient pertains only to its individual use – and a contrary practice is neither honourable nor honest.

CHAPTER IV. – "MEDICAL" ETIQUETTE, OR THE RULE OF THE PROFESSION ON COMMENCING PRACTICE, ETC.

1. In the absence of any published rule, or collegiate instruction, on such and kindred matters, it is not to be wondered at that young practitioners should be so generally ignorant of the "unwritten" custom or etiquette (diverse as it is from that pursued in ordinary social life, in relation to new residents) expected from members of the profession on commencing or changing the locality of practice, in town or country, – and which entails on each newcomer, young or old, an obligation to call, with as little delay as may be, upon every duly qualified, legitimate medical practitioner resident within a reasonable distance of his own selected place of abode, and courteously announce his intention to practise in the locality.

NOTES

[1] As a guide to young practitioners, it may be well to note that, in consultations, it is customary for the family doctor to precede the Consultant into the sick-room, and to retire therefrom after him. 'Tis scarcely necessary to add that, as a rule, it rests with the Consultant, and not with the regular attendant, to fix the hour of meeting.

[2] What, it has been critically asked by an eminent practitioner since the preceding was penned, should be the conduct of the Consultant when he finds that the ordinary medical attendant has misunderstood the case, or, it may be has committed a grievous error. In obedience to the "royal law" . . . [?].

[3] The development of a tariff of fees which shall be acknowledged by the profession as com-

pensative, and, by the public, as reasonable, must, it is to be feared, be regarded as utopian, so long as the medical and surgical professions hesitate to found their claim to remuneration upon the value of their time and skill, and persist in the objectionable system of "drug payment."

CHAPTER 8

M. ANNE CROWTHER

FORENSIC MEDICINE AND MEDICAL ETHICS IN NINETEENTH-CENTURY BRITAIN

Where did British medical students of the past century learn about medical ethics? Probably not from regular reading of Thomas Percival or any of the other essayists on the subject, though reprinted editions of Percival, some shorn of his "old fashioned" concerns, affirmed his continuing appeal [25]. This, however, may be exaggerated. Percival was probably more famous in his lifetime for his *A Father's Instruction to his Children, consisting of Tables, Fables and Reflections, Designed to Promote the Love of Virtue, etc.* The work ran to ten editions between the 1770s and 1800. *Medical Ethics; or a Code of Institutes and Precepts, adapted to the professional Conduct of Physicians and Surgeons*, reached a third edition only in 1849, though it was also printed in two editions as part of Percival's collected essays. It would not appear to have been available for British students to buy in the mid-nineteenth-century. As Burns has pointed out, "no British physician between 1803 and 1850 had written a monograph on medical ethics," and the revival of interest in 1850 was largely due to American influences ([6], p. 304).

Medical ethics never gained a natural home in the curriculum of medical schools, which relied on a few set textbooks, and much of the formal instruction was imparted through detailed lectures given at dictation speed. Ethical questions may have appeared in informal discussion, or through the precept and practice of clinical teachers, but the most systematic instruction came in lectures and standard textbooks on medical jurisprudence, which set out the legal obligations of the medical profession while dealing incidentally with wider ethical issues. Placing medical ethics in this practical and legalistic framework was to distance it from the intricate professional, moral and social concerns of Gregory ([17]) or Percival; instead, the student was offered a body of

R. Baker (ed.), The Codification of Medical Morality, 173–190.
© 1995 *Kluwer Academic Publishers. Printed in the Netherlands.*

case law to clarify the relations between doctor and patient, including such issues as professional secrecy and the legal duties of medical men. Alongside the discussion of legal matters, however, was another agenda of medical ethics which rests on assumptions and consensus rather than legal precedent. It concerned the doctor's social responsibilities, and his attitudes towards actual or suspected criminals and their victims.

The ethical function of medical jurisprudence long survived in British medical schools. In 1984, the student looking for information on the doctor's obligations to his patient, criminal negligence, the legal liability of the National Health Service, mistaken diagnosis, and informed consent, would find it in the fourth chapter of *Taylor's Principles and Practice of Medical Jurisprudence*, sandwiched between a chapter entitled "Death" and another on "Medico-legal examination of the living" ([34]). This was the thirteenth edition of a standard work dating from 1844. The first edition, like earlier nineteenth-century texts on this subject, was greatly concerned with the analysis of poisons, interpretation of wounds, and other conventional subjects, and its formal instructions on medical ethics were brief: they included the doctor's duties in detecting crime, diagnosing insanity and malingering, and giving evidence competently in court. Successive editions of Taylor, and other texts on forensic medicine, chart the growing complexity of the law in relation to medicine, and reveal changing social expectations of the profession's powers and responsibilities.

It is hardly surprising that medical ethics were, and are, presented to the student in a legal context, since most of the teaching on the subject has aimed to instruct future doctors and lawyers in ethical issues which can be tested in court. In each country, the law provides the framework of medical ethics on many questions, including the legality of euthanasia and abortion and the treatment of severely handicapped infants. Each legal system has its own history of individual cases which have set the boundaries of ethical debate. As far as British students were concerned, formal discussion of medical ethics was long buried in the textbooks among the more practical material on toxicology and diagnosis of physical violence. Its status in medical education generally was weak. Textbooks like Taylor's, although undoubtedly widely consulted and frequently reprinted, were not usually a compulsory part of the examinations in English medical schools, though attendance at a course of lectures on medical jurisprudence might be required. The Scottish schools, however, introduced a compulsory element of medical jurisprudence for students in both medicine and law early in the nineteenth century, and this remained until after the second World War ([12], pp. 7, 99–100). Nevertheless, medical ethics remained a marginal part of a marginal subject.

In any discussion of medical ethics, the name of Thomas Percival is unavoidable. Recent writings on Percival have questioned whether he was writing about ethics, rather narrowly defined as duties towards the patient; or professional etiquette, the formal relationships between members of the profession ([3], [35]). To Percival, this distinction would have made little sense: he offered his theory of professional relations and hospital management as a way of safeguarding patients' interests, as well as maintaining the respectability of the profession. In this debate the legal dimension of his most quoted work should not be forgotten, since it clearly shows the overlapping of professional etiquette with much wider ethical issues. The first version of *Medical Ethics* was privately printed in 1794 under the title *Medical Jurisprudence*, and Percival, unlike Gregory, showed much interest in the legal role of the practitioner. Waddington tends to shrug off Percival's chapter on medical jurisprudence as though it were not an integral part of his views on medical ethics, but this distorts the book's purpose ([36], p. 157). Percival's definition of medical ethics was not confined to relations within the profession or duties towards patients, but embraced the practitioner's whole responsibility to society. Serving the ends of justice was a crucial part of this responsibility. The doctor had obligations to colleagues and patients; but in the end, Percival reminded him, his duty to justice and his country should overcome everything else, especially as his privileged position exempted him from other duties such as military service. The medical man might well find himself in an uncomfortable role, when asked by his neighbours to exempt them from their own civic or military obligations:

> No fear of giving umbrage, no view to present or future emolument, nor any motives of friendship, should incite to a false, or even dubious declaration. For the general weal requires that every individual, who is properly qualified, should deem himself obliged to execute, when legally called upon, the juridical and municipal employments of the body politic. And to be accessary, by untruth or prevarication, to the evasion of this duty, is at once a high misdemeanour against social order, and a breach of moral and professional honour ([25], Chapter 2, Article 20).

In the chapter entitled "Of Professional Duties in certain Cases which require a Knowledge of Law," Percival argued that medical men needed some legal knowledge in order to assist justice. He was not concerned with forensic pathology in its modern sense, but with laws which the profession must understand in order to deal correctly with such matters as homicide and infanticide, rape, insanity, disputed wills and sanitation. Some of his discussion appears on the surface to be concerned only with professional etiquette: like nearly every subsequent British writer on the subject, Percival was anxious lest the adversarial system of law in Britain should lead the profession into futile courtroom battles which exposed its ignorance and invited public ridicule. He therefore deplored the spectacle of medical men attempting to discredit each other's evidence. But his emphasis that medical evidence should be rooted in humility and scientific enquiry was as much in the interests of justice as of the medical profession. He also considered more straightforward ethical issues: doctors ought not to resist unpaid service at coroners' inquests, and should not be mealy-mouthed as medical witnesses, even if the result were the death penalty for the accused ([25], Chapter 4, Articles 8–9). He entered some of the murkier areas of medical ethics, including medical attendance at duels, and questions of malpraxis.

Percival's medical ethics were therefore rooted in the law, which gave the medical man a special status. Under the rules of evidence, witnesses could speak only of fact, not of conjecture: the medical witness alone was permitted to conjecture on many crucial subjects such as the cause of death or the signs of recent childbirth. Percival argued that the medical man could not shirk legal duties, a view which undermines the mercenary interpretation sometimes applied to his medical ethics. The medical witness was unpaid, and his courtroom appearance might endanger his reputation; nevertheless, his ethical responsibility was clear:

> It is a complaint made by coroners, magistrates, and judges, that medical gentlemen are often reluctant in the performance of the offices required from them as citizens qualified, by professional knowledge, to aid the execution of public justice. These offices, it must be confessed, are generally painful, always inconvenient, and occasion an interruption to business, of a nature not easily appreciated or compensated. But as they admit of no substitution, they are to be regarded as appropriate debts to the community, which neither equity nor patriotism will allow to be cancelled ([25], Chapter 4, Article 18).

Percival's approach to medical ethics was not particularly novel, but was rooted, like Gregory's, in the claim for medicine as both a liberal and a scientific profession. Such a view became the common currency of medical thought in the late eighteenth century, not least in Edinburgh, where medical men gave Enlightenment philosophy a practical application ([7], pp. 124–94; [19], [28], pp. 1–3). The case-notes and medical theories of the learned Dr. Percival were warmly received in Edinburgh by the elder Andrew Duncan, and frequently appeared in Duncan's periodical, *Annals of Medicine*, precursor of the *Edinburgh Medical Journal*. Duncan's own pioneering lectures on medical jurisprudence during the 1790s made similarly strong claims for the importance of medicine in upholding justice. His list of the legal responsibilities of medical men, as outlined in his proposals for a new chair of Medical Jurisprudence and Medical Police, hardly differed from Percival's, and were rooted in European theories of medical police ([14]).[1] Here the duties of the doctor, far from being defined only in terms of his patients, encompassed the whole fabric of society, from public health to the administration of justice.

Although Percival's work will doubtless continue to attract new interpreters, its immediate legacy was to link medical ethics with the needs of the law, and so bring it within the province of the teachers of forensic medicine. In the course of the nineteenth century, the precepts of Percival and Duncan were expanded in the forensic textbooks and lectures of others, to take account of new legal duties of the profession. Life insurance practices, registration of deaths, the sale of drugs and poisons, workmen's compensation for industrial injuries, and many other legislative developments had to be explained to medical students. After the 1858 Medical Act and the establishment of the General Medical Council, the GMC's legal power to decide what constituted "infamous conduct in a professional respect" had also to be conveyed to new practitioners. Its definition of "infamous conduct" ranged from criminal offences to advertising: it upheld both a moral code and a professional etiquette. Students of medical jurisprudence would find these issues side by side in the textbooks, placing medical ethics in a legalistic framework ([16], Chapters 1–2).

The earlier British writers on forensic medicine, following Percival's example, were chiefly concerned to reinforce one basic ethical point: the duty of the medical man to give expert evidence in court, and in a manner reflecting credit on his profession. The substantial text on medical jurisprudence by Paris and Fonblanque published in 1823 is representative of this approach:

> Let the physician then, who approaches the tribunal of justice in order that he may promote by his science the due execution of the laws, fully appreciate the heavy responsibility of his situation; let his evidence be so distinguished by its dispassionate and inflexible character, and his opinions be so matured by study and fortified by experiment, as not only to ensure for himself the respectful attention of the court, but to afford a practical illustration of the just pretensions and importance of the liberal profession which he represents ([24], p. 399).

Just as Percival took the opportunity to encapsulate in his medical ethics very specific references to infirmary management in Manchester, so John Ayrton Paris, a popular lecturer on Materia Medica, and a leading member of the Royal College of Physicians, used his text to defend the College's ancient privileges over medical licenses in London ([24], Vol. 1). Local concerns were fused with the highest rhetoric on wider issues, especially urging the profession to act as expert witnesses in spite of the two major disincentives: lack of fees, and the fear of making fools of themselves. In England, the coroners' courts were unpopular for arbitrarily summoning unpaid medical witnesses, until the Medical Witnesses Act of 1836 set up a schedule of fees, but the work was still felt to be underpaid. In Scotland, where autopsies and prosecutions were both ordered by the Procurators Fiscal, medical witnesses were paid, and Robert Christison, the leading figure in Scottish forensic medicine, continued to urge compulsory education in forensic medicine to prevent medical men from exposing the profession to ridicule by their performance in the witness box [10].

Like Percival, the early writers on forensic medicine made no particular distinction between medical ethics and medical etiquette, since etiquette in the courtroom was assumed to protect the interests of society as a whole. Indeed, the ends of justice could not be served unless a proper etiquette were established. The fear, frequently expressed in the early writers, was that medical men would be set against one another by the lawyers, thus bringing the profession into disrepute. John Gordon Smith, then a lecturer on medical jurisprudence in London, drew an affecting picture of the unimpressive medical witness, enduring

> the scrutiny and displeasure of the bench, the brow-beating of the bar, the scorn, laughter, or contempt of the audience, the discontent of his friends, and the exposure of the public press, with all the consequences that may follow to his reputation and fortune ([32], p. 41).

Medical jurists often justified the importance of their subject by arguing that a proper study of forensic medicine would carry normal medical etiquette into the courts. If medical men were reasonably educated in the principles of forensic medicine, unseemly court-room battles would be avoided, and the ends of justice better served ([23], p. 25; [31], p. 9). In these terms, etiquette served a higher ethical purpose; but, in practice, the law could expose conflicts between medical ethics and conventional medical etiquette.

A principle apparently established in law was that the medical man had no power to withhold evidence in a criminal court ([24], Vol. 1, p. 160). The test case was the Duchess of Kingston's bigamy trial in 1776, where her medical attendant refused to give evidence about her obstetric history until the judge ruled that a practitioner's duty to the law overrode the confidentiality of the patient ([16], p. 56). But this apparently straightforward maxim had its complications, and was not universally accepted by the profession: John Gordon Smith, for example, argued that doctors should have the same privilege of silence as lawyers ([32], p. 97). In any case, giving evidence when called as a witness was one matter; informing the authorities when the practitioner suspected his patient to be the victim of foul play was another. Two well-known trials provoked intense public debate on the ethics of professional etiquette. The first, the Wooler "slow poisoning" case of 1855, led to the acquittal of Joseph Snaith Wooler, a retired merchant, from a charge of poisoning his wife Jane with arsenic ([1], pp. 375–90). Mrs. Wooler's medical attendant, Dr. T. Hayes Jackson, suspecting poison, called in a second doctor, who also became alarmed; but they lacked the knowledge or resources to carry out adequate tests. The Woolers lived in Burdon, near Darlington, and there was no chemical analyst in the district. The two practitioners sent specimens of urine for analysis to Christison in Edinburgh, but in the ten days between first discussing the case and receiving an answer from Christison confirming the presence of arsenic, the unfortunate Mrs. Wooler died.

Mrs. Wooler's two attendants were excoriated by the prosecution, the judge and the press for failing to inform any authorities of their suspicions [11]. It appeared that they had put the claims of confidentiality to the client (in this case, Mr. Wooler, as the fee-payer, rather than his wife) before those of the law. But the medical profession closed ranks around them. The medical press argued that few medical men would endanger their livelihood by voicing suspicions when there was no positive proof.[2] The prosecution believed that the doctors had an ethical duty to protect Mrs. Wooler once they suspected poison, but the issues in this case were by no means clear.

Christison, like other writers on the subject, distinguished between "scientific" and "moral" evidence. Medical men were responsible for offering whatever scientific evidence was available, but the court would also take account of "moral" evidence, by which was meant the general circumstances of the crime, the testimony of witnesses, etc. Mrs. Wooler's attendants were unsure of their scientific grounds, nor had they any "moral" evidence, in view of the apparently harmonious Wooler household: indeed, Wooler's acquittal was partly due to the prosecution's failure to offer any motive for the crime. In the circumstances, a medical man might well consider his ethical duty uncertain, and the professional consequences of a wrongful allegation far more damaging. Christison and other correspondents in the *Edinburgh Medical Journal* sympathized with this professional dilemma, since it placed the medical attendants in an impossible situation. Christison's own view was that Mrs. Wooler, if no-one else, should have been warned, but he acknowledged the predicament: if medical men failed to recognise the symptoms of poisoning, they were seen as incompetent; if they made allegations without suitable proof, they were open to the law of libel and professional ruin.

The tragedy ended in farce when Dr. Hayes Jackson sued Wooler for unpaid medical bills: forty journeys to see Mrs. Wooler, at three shillings and sixpence each. Wooler's defence was that Jackson was incompetent, a somewhat ironic argument, for if Jackson had possessed the skill of a Christison, he might have given stronger evidence against Wooler. The court found against Wooler, and the crowd, apparently convinced of his guilt, drew Jackson's carriage in triumph through the streets. The medical reaction to the case did not dispute that there was an ethical code which transcended professional secrecy, but also indicated the enormous difficulties of putting this code into practice in doubtful cases.

An even more notorious case in 1865 provoked a public examination of the distinction between medical ethics and etiquette. This was the trial of William Pritchard for the murder by poisoning of his wife and mother-in-law. On three occasions, Pritchard, himself a medical man, called in Dr. James Paterson as a second opinion. Paterson afterwards claimed that he had immediately suspected poisoning but, when Pritchard's mother-in-law died, did nothing to alert the authorities apart from refusing to sign the death certificate: Pritchard signed it himself. Paterson wrote an ambiguous letter to the Registrar, which was interpreted as refusal to sign because the dead woman was not his own patient. Unwisely, Paterson seems to have been more worried about appearing incompetent rather than unethical, since he stuck to his claim that he had correctly diagnosed the symptoms of poison, even though they were by no means straightforward. In court, Paterson therefore had to defend his conduct on the grounds of professional etiquette: that he had been called in purely as a consultant, and Pritchard's family was not his responsibility. This stirred the judge into a vigorous assault on medical etiquette:

> I care not for professional etiquette or professional rule. There is a rule of life and a consideration that is far higher than those – and that is, the duty that every right-minded man owed to his neighbour, to prevent the destruction of human life in this world, and in that duty I cannot but say Dr. Paterson failed ([29], p. 283).

The medical press was at first inclined to accept this comment. The *Lancet*'s editorial argued that Paterson's views of medical etiquette were unfounded, for no rule of etiquette could override a doctor's duty to protect his patient. It concluded:

> He has done a . . . grievous wrong to the whole profession in throwing the burden of the blame due to his cowardice and want of judgment upon an alleged etiquette which does not exist [*Lancet* 15 July 1865, p. 70].

But, as in the Wooler case, professional opinion divided. The *BMJ* at first agreed with the judge's comments, and criticized Paterson for not having chemical tests carried out as soon as he suspected poison: "his first duty was to attempt to save his patient, although doing so might apparently cause a slight breach of professional etiquette." Once Paterson had become the butt of the press, especially the *Times* (scourge of professional pretensions), the *BMJ* rallied in his defence, drawing a clear distinction between the claims of ethics and etiquette, and arguing that Paterson's "error" should not be used to berate the whole profession:

> *The Times* has taken up the position that Mrs. Pritchard was sacrificed to etiquette, and has, in fact, drawn from this incident the moral, that the medical profession are constantly in the habit of sacrificing their patients at the shrine of their favourite household goddess, etiquette.

Once again, the practical difficulties facing a medical man in this position were accepted as an extenuation, for

> the difficulty of believing, and of diagnosing, such a horrible act of criminality . . . , especially as being committed by a brother medical man, must of necessity be very great.

The *BMJ* also printed a fervent letter in support of Paterson, which claimed that "this note of praise is merely one sound in a chorus lifted up by many in approval of Dr. Paterson."[3] In all these examples, however, the medical response to Paterson's dilemma showed no confusion over ethics and etiquette. No one attempted to argue that consideration for a fellow practitioner overcame the doctor's duty to his patient. Support for Paterson, as for Jackson, was based on their practical difficulties in making an appropriate decision when diagnosis was uncertain. The ethical issues were essentially as voiced by Percival in his chapter on medical jurisprudence: practitioners must observe scrupulous etiquette in relation to one another, but the claims of society, and of justice, overrode professional relationships.

Suspected poisoning became a simpler matter by the end of the nineteenth century because practices in health care for the upper classes changed. In John Glaister's textbook, published at the turn of the century, the chapter on professional secrecy offered straightforward advice: "Have the patient removed to hospital or nursing home" ([16], p. 643). Yet even if poisoning were a very rare crime, the reaction to it had implications for the later dilemma over child abuse, where the practitioner was urged to adopt similar tactics – removal of the putative victim to a pediatric unit. The dilemma of Jackson and Paterson is not far removed from recent British cases involving child sexual abuse, where diagnosis is uncertain and doctors are excoriated if a child comes to harm, but in danger of professional ruin if they intervene "officiously" to separate children from parents.

The issue was not a minor one, since it affected any practitioner who suspected that a crime had taken place, including cases far more common than poisoning. It recurred particularly in cases of illegal abortion. Here, however, the question was not one of professional etiquette versus the law, but rather that the demands of the law did not accommodate well to medical ethics. This was the mirror-image of the poisoning problem, in that the doctor might consider that he was protecting his patient's interests if he concealed, rather than exposed, the crime.

John Keown has argued that British abortion law, as it developed in the nineteenth century, showed the strong influence of medical men, anxious both to protect lives from an operation perceived as dangerous, and to establish the respectability of their own profession ([18], Chapter 2). Nevertheless, the problem of a doctor confronted with the aftermath of an abortion remained a murky area of medical ethics, even if the operation had obviously been performed by a professional abortionist. To alert the police in such cases would ruin the reputation of the patient, and probably discourage other women from applying for medical help until it was too late. Nor would the doctor relish the prospect of appearing in court as a witness.

As late as 1922, the Lord Chancellor, the Earl of Birkenhead, published what was perceived as a "definitive" essay on the subject of medical secrecy, under the title "Should a Doctor Tell?" ([5], pp. 52–5). He enunciated firmly the doctor's duty to alert the authorities. However, it is clear from the correspondence in the Lord Chancellor's Office, that his action had been provoked by a considerable conflict over medical secrecy, and, amongst other issues, by dissatisfaction amongst the police about the number of cases which came to light only after a woman had died while receiving regular medical treatment [27]. The woman might even have told her practitioner the name of the abortionist, but this information did not constitute a legally sound "dying deposition," and could not be used in evidence. Evidently, an even larger number of non-fatal cases must have gone unreported by the profession. Sometimes, of course, the abortionist was a fellow-practitioner, but medical etiquette did not enter into the question. All treatises on the subject spelt out the doctor's legal duty, yet it was frequently evaded.

Although the textbooks stressed that abortion performed on another person was a serious offence, they were less clear about a woman's criminal liability if she attempted an abortion on herself. Here the doctor might be justified in keeping his patient's counsel. Glaister's textbook reflected this confusion by citing a variety of conflicting legal opinions, and he ended by quoting a judge's comment from 1896:

> If a poor, wretched woman committed an offence for the purpose of getting rid of that with which she was pregnant, and of saving her character, her reputation, and, it might be, her very means of livelihood, and if a doctor was called into assist her – not in procuring abortion, for that in itself was a crime – but called in for the purpose of attending her and giving her medical advice . . . he [the judge] doubted very, very, very much whether he would be justified in going forth and [alerting] the Public Prosecutor . . . ([16], p. 61).

The criminal law therefore raised a number of problems concerning professional secrecy, but the doctor's obligations under the civil law were even more uncertain ([16], pp. 57–8). In 1900, an English judge decided that a doctor could not be compelled to give evidence in a civil suit if he would then be liable to be sued by his patient, but contradictory decisions could also be found. ([37], p.138)

Glaister's approach to the central issues of medical secrecy was fairly typical of the British approach, both in lectures and textbooks on medical jurisprudence, being couched in legal rather than ethical terms. Percival's interest in moral questions had been echoed, to some extent, by the writers on medical jurisprudence who were closest to him in time, such as John Gordon Smith and John Ayrton Paris; by contrast standard works from the mid century onwards were more concerned with ethics as defined by statute or legal opinion. Formal medical ethics became a legal rather than a moral code, and Glaister's pragmatic comments reflected this.

The development of forensic medicine as a specialty produced its own problems of medical etiquette. Earlier texts offered relatively little specialized information, and stressed that every practitioner should master the elements of forensic medicine. Rapid progress in toxicology, and the belief that experience in morbid anatomy counted for more than general education, began to move medical jurisprudence from the domain of the ordinary practitioner to the expert. The Wooler case itself showed that ordinary practitioners could not handle complicated chemical analysis. The Scottish system, where the Procurators Fiscal encouraged the development of a small cadre of recognised medical experts, revealed some of the problems. Christison's own work was full of contradictions: on one hand he called successfully for medical jurisprudence to be a compulsory part of the medical examinations in Edinburgh, and hence part of the equipment of all medical graduates; on the other, his own research took him into complex areas of toxicology where few general practitioners could hope to follow.

Scottish legal practice also created tensions because it interfered with notions of medical etiquette. The litigious surgeon James Syme protested vigorously when a police surgeon entered his hospital ward to take a "dying deposition" from a patient who then recovered: to Syme, this was an unwarranted interference of the law between the patient and his doctor [2]. In 1853, the Royal College of Physicians of Edinburgh, supporting Syme, petitioned the Lord Advocate to ensure that the "physician of first attendance," should not be superseded by the Crown's medical expert [30], but the question was largely left to individual discretion. One of Christison's successors in the chair of medical jurisprudence, Henry Littlejohn, advised his students that, if they were ever summoned to act for the authorities, they should be sure to consult local doctors, to avoid ruffling their feelings [20]. Intrusion by an outside expert raised not only questions of propriety, but the possibility that he would deprive local practitioners of fees for the autopsy and court appearances. The challenge to medical etiquette of this interposition between a patient and his usual doctor was justified in terms of a higher claim – expert knowledge as an instrument of the law.

These formal areas of forensic medicine were underlaid with other ethical considerations taken for granted by its teachers. By its very nature, forensic medicine invited social commentary, and provided well-worn paths through moral dilemmas. Here one could include attitudes towards homosexuality, infertility, insanity, and a variety of other subjects in which medical, legal and ethical issues were intertwined. Two examples should suffice: infanticide and rape.

In the eighteenth century, a woman accused of child murder had to prove that the child had been born dead or died of natural causes: in the absence of such evidence, she was automatically considered guilty. This reversed the normal procedure, where the onus of proof lay on the prosecution. An Act of 1803 in England (1809 in Scotland) liberalised the law: the court now had to prove that a murder had been committed [4].[4] If this were not possible, the accused might be found guilty of lesser crimes; in England of concealing the birth, in Scotland of concealing pregnancy. Both carried a maximum sentence of two years' imprisonment.

Throughout the nineteenth century, the law in cases of child murder caused a conflict in medical ethics, and medical students might have been pardonably confused had they consulted a number of the most respected texts. Not only did the law run contrary to the ethical ideas of many medical practitioners, but there was apparently no agreement on what those ethics should be. In effect, the law offered only two sentences for those accused of child murder: capital punishment or up to two years' imprisonment. This led, in Taylor's opinion, to excessive leniency in sentencing, even in obvious cases of murder ([33], pp. 498–9). The role of the medical man was crucial, since only he could confirm, in the absence of other witnesses, whether a child had been born alive. This was the problem. After a number of difficult cases, judicial precedent established that child murder was possible only if a child were born alive: but the judicial and the medical definitions of a live birth differed. In the practice of the courts, a live birth meant that a child must be fully separated from its mother: if it were strangled or otherwise killed during the process of birth, murder could not be proved. For medical experts, this simply showed that the law was an ass, since, medically speaking, the child was alive as soon as it began to breathe, regardless of whether it were fully separated from its mother ([24], Vol. iii, p. 90; [33], pp. 471, 474–5; [16], pp. 488–91). Cases where the infant showed marks of extreme violence to the head could still lead to acquittal or a short sentence if live birth in the legal sense could not be proved.

Textbooks throughout the century naturally devoted a great deal of attention to the physical proofs of a live birth, and betray an obvious impatience with the law's ability to confound the unwary medical witness. The root of the problem, acknowledged more or less openly, was William Hunter's influential essay, *On the Uncertainty of the Signs of Murder, in the case of Bastard Children*, published in 1783. Hunter had helped to overturn the eighteenth-century law on infanticide by casting doubt on the hydrostatic test for live birth. An infant's lungs, Hunter argued, might well float in water for a number of reasons, including putrefaction, without proving that it had been born alive.

Nineteenth-century authorities disputed Hunter's findings, since air in the lungs was unlikely to result from putrefaction if this were not apparent elsewhere. None of the alternative efforts, including Ploquet's attempts to determine a live birth by the weight of the lungs, was considered as sound as the hydrostatic test in offering prima facie evidence ([24], Vol. 3, p. 117; [31], p. 340; [23], p. 275). Yet Hunter's legacy was to make courts extremely suspicious of the hydrostatic test, and any medical man unaware of the complex debates on the subject, or unable to present his views coherently, might find his evidence disregarded.

In fact, the durability of Hunter's argument lay less in his scientific attack on the hydrostatic test than in his general attitude towards women accused of this crime. He could not accept the law's severity towards the kind of woman usually involved in such cases, poor, friendless and betrayed. John Gordon Smith, although a believer in the hydrostatic test, was lengthily sentimental towards errant mothers, urging medical witnesses to take all extenuating circumstances into account ([31], pp. 340–75). Such sentiments were readily adopted by Victorians brought up on *The Heart of Midlothian* and *Adam Bede*, but had strong opponents also. Percival, who quoted Hunter at length on the motives for infanticide, argued that Hunter had "exalted the sense of shame into the principle of virtue," and had "mistaken the great end of penal law, which is not vengeance, but the prevention of crime," though Percival agreed that the law should be amended to give more appropriate punishments ([25], Chapter 4, Article 11). Later writers also objected that the crime was masked by popular sentimentality, as lampooned by Alexander Ogston, future professor of medical jurisprudence at Aberdeen:

> It has been deemed unfair, if not unjust, to doom to the same infamy and the same punishment the hardened and atrocious criminal and the young female of character and reputable connections who, betrayed by the arts of the base seducer, to avoid shame and disgrace and the ruin of all her earthly prospects, is driven by a momentary impulse either to stifle her new-born infant, or to imbue her hands in its blood at its birth. . . .
>
> It is not for me to dwell on the flimsy character of the pleas set up, in our courts of law, in mitigation of the offence . . . particularly as these are beginning to be acknowledged on the part of legislators . . . ([23], pp. 207–8).

The lack of interest in puerperal mania shown by the forensic experts is noteworthy, since they always considered the diagnosis of insanity as part of their brief. Paris and Fonblanque did suggest the possibility of puerperal mania ([24], Vol. 3, p. 129), but this was not taken up in Taylor's more influential work. In his first edition, Taylor briefly mentioned the possibility of mania, but his description implied that such delirium rarely occurred before the third day after birth, and often not for a fortnight; that those who destroyed their children under this affliction usually lacked all motive to do so, and that they never attempted to conceal the crime ([33], p. 654). His 1865 edition, after a couple of decades of great controversy over infanticide, expanded this argument, but did not alter it. By definition, therefore, a woman who destroyed her child at birth, who had a *social* motive for doing so, and who concealed her crime, could not be suffering from a mental condition.

In spite of all pressure to change the law, little was done until the Infanticide Act of 1922 fixed the concept of puerperal insanity more firmly into English law, and the crime was associated with manslaughter, subject to wide discretion on sentencing, rather than murder.

Attitudes towards rape victims provide another example of medical practices which would today attract a rather different ethical interpretation ([26],

p. 220 ff.). Unlike infanticide, there was little disagreement among the authors either on the ambiguous physical signs of rape, or in attitudes to the victim. Percival was sympathetic to the rape victim, as having lost "the most sacred of all personal property," but he then fell into a common pattern where more effort was given to discrediting a rape charge than substantiating it ([25], Chapter 4, Article 16). From Duncan's lectures in 1800, students learned that rape could not be committed on an able-bodied woman, unless she had fainted or was overcome by drink [13]. In 1823 Paris and Fonblanque wrote:

It is at all times difficult to believe that in a mere conflict of strength, any woman of moderate power of body and mind, could suffer violation, so long at least as she retained her self possession ([24], Vol. 1, p. 423).

In Christison's lectures of 1831, this appears in his student's notes as;

A woman possessed of her senses & natural strength can resist a rape – the weight of authority seems to be against the idea that a woman's *will* may be forced; – for if you admit it, how are you to draw the boundary between forcing the will & seduction [9].

Taylor accepted that rape of a healthy woman was possible, but only if she were overcome by narcotics, drink, terror or exhaustion; he also prefaced his discussion by noting Amos's comment from the 1830s that "for one real rape tried on the Circuits, there were on the average twelve pretended cases!" ([33], pp. 574–9). These views were echoed in 1878 by Ogston, who repeatedly emphasized the likelihood of false charges of rape being brought either by immoral women, or by those who regretted having given their consent to intercourse ([23], pp. 119–23). In 1893, Dixon Mann in Manchester added a class dimension; that rape of a healthy woman was more likely to occur in the refined classes;

Women of the lower classes are accustomed to rough play with individuals both of their own and of the opposite sex, and thus acquire the habit of defending themselves against sportive violence. In the majority of cases such a capacity for defence would enable a desperate woman to frustrate the attempts of her intentioned ravisher. A delicately nurtured woman, on the other hand, is so appalled by the unwonted violence that her faculties may be partially benumbed, and her powers of resistance correspondingly enfeebled ([22], p. 96).

Shortly afterwards, Harvey Littlejohn, who had succeeded his father as Professor of Medical Jurisprudence in Edinburgh, was telling his students that poor women and farm girls should be expected to have more presence of mind than women of higher social status, but "Generally speaking a single woman can resist a single man" [21].

Nearly all the British authorities gave careful consideration to this topic in moral as well as medical terms, and the example serves to show how certain comments became fossilized in textbooks and lectures over long periods of time. Since forensic medicine relied heavily on examples from case law, the same cases and commentary were constantly repeated. By the nature of the subject, medical jurists were drawn into consideration of the "moral" evidence, such as assumptions about the nature of rape. These generalizations were handed down through generations of students, often barely altered from one authority to another. The development of conventional masculine wisdom about rape, whether medical or legal, appears to have been closely mirrored in the courts, where accusations of rape were extraordinarily difficult to sustain, and a large proportion were not even brought to trial ([8], pp. 82–4).

By the end of the nineteenth century, forensic medicine had become a rag-bag subject, containing the details of morbid anatomy, the clinical work of the police surgeon, and the increasingly complex chemical analysis required for toxicology and serology. Inevitably, the more complicated parts of the subject were being separated from forensic medicine, and became the prerogative of scientists rather than medical men. This left the forensic textbook with a residue, not only of morbid anatomy and clinical investigation, but a slowly accumulated collection of medical ethics, sometimes reinforced by legal requirements, sometimes by convention.

Percival's medical ethics were presented in a framework of medical jurisprudence. As a man of the Enlightenment, the essential harmony between science and justice, exemplified in the doctor's legal role, appealed strongly to him. John Gordon Smith's long homilies on morality and forensic medicine, while lacking Percival's intellectual coherence, nevertheless echoed his concerns. Later writers, like Taylor or Glaister, being mainly concerned with the scientific detail of medical jurisprudence, tended to deal with the subject of medical ethics simply in terms of the law's demands on the medical profession. In the teaching of forensic medicine, medical ethics appeared as a blend of legal decisions, professional etiquette, and social philosophy: and its exponents would probably have been surprised at the efforts of modern writers to disentangle them. In this unsystematic fashion, the student acquainted himself with some of the basic principles of his profession, and, we might argue, acquired in the process a body of folklore difficult to shake.

NOTES

[1] There is an excellent list of the type of Continental writings on medical jurisprudence available to the educated medical man of the time in the first chapters of [24].
[2] For further accounts, see the *Lancet*, 18 March 1856, p. 274, and 26 April 1856, p. 470.
[3] For a full account, see *BMJ* 15 July 1865, p. 50; 22 July, pp. 63–4; 29 July, pp. 102–3, and [15], p. 164.
[4] 43 Geo. III c. 58, and 49 Geo. III c. 14.

BIBLIOGRAPHY

1. *Annual Register*, 1855.
2. Anon. (J. Syme): 1855, *Illustrations of Medical Evidence and Trial by Jury in Scotland*, Sutherland & Knox, Edinburgh.
3. Baker, R.: 1993, "Deciphering Percival's Code," in Baker, R., Porter, D., and Porter, R. (eds.), *The Codification of Medical Morality: Historical and Philosophical Studies of the Formalization of Western Medical Morality in the Eighteenth and Nineteenth Centuries: Volume One: Medical Ethics and Etiquette in the Eighteenth Century*, pp. 179–211, Kluwer Academic Publishers, Dordrecht.
4. Behlmer, G. K.: 1979, "Deadly Motherhood: Infanticide and Medical Opinion in Mid-Victorian England," *Journal of the History of Medicine and Allied Sciences* 34(4), 403–27.
5. Birkenhead, Earl of (F. E. Smith): 1922, "Should a Doctor Tell?" in *Points of View*, Vol. 1, pp. 33–76, Hodder & Stoughton, London.
6. Burns, C.: 1977, "Reciprocity in the Development of Anglo-American Medical Ethics, 1765–1865," in Burns, C. (ed.), *Legacies in Ethics and Medicine*, pp. 300–6, Science History Publications, New York; this volume, pp. 000–000.
7. Chitnis, A.: 1976, *The Scottish Enlightenment*, Croom Helm, London.
8. Conley, C. A.: 1991, *The Unwritten Law: Criminal Justice in Victorian Kent*, Oxford University Press, New York.
9. Christison, R.: MS Notes of his Lectures, MS 2958, Wellcome Library, London.
10. Christison, R.: Nov. 1851, "On the Present State of Medical Evidence," *Edinburgh Monthly Journal of Medical Science* 13, 401–30.
11. Christison, R.: 1856, "Account of a Late Remarkable Trial for Poisoning with Arsenic," *Edinburgh Medical Journal* 1, 625–32; 707–18; 759–61.
12. Crowther, M. A. and B. M. White,: 1988, *On Soul and Conscience: the Medical Expert and Crime*, Aberdeen University Press, Aberdeen.
13. Duncan, A.: n.d., MS Lectures on Medical Jurisprudence, Dc.8.158., Edinburgh University Library, Edinburgh.
14. Duncan, A.: n.d., "A Short View of the Extent and Importance of Medical Jurisprudence, Considered as a Branch of Education," (special collections), Edinburgh University Library, Edinburgh.
15. Glaister, J.: Sept. 1886, "Medico-Legal Risks Encountered by Medical Practitioners in the Practice of their Pofession," *Glasgow Medical Journal* 26(3), 161–80.
16. Glaister, J.: 1921, *A Textbook of Medical Jurisprudence and Toxicology*, (1st ed., 1902), 4th ed., Livingston, Edinburgh.
17. Gregory, J.: 1772, *Lectures on the Duties and Qualifications of a Physician*, 2nd ed., W.

Strahan, London.
18. Keown, J.: 1988, *Abortion, Doctors and the Law. Some Aspects of the Legal Regulation in England from 1803 to 1982*, Cambridge University Press, Cambridge.
19. Lawrence, C.: 1985, "Ornate Physicians and Learned Artisans: Edinburgh Medical Men, 1726–1776," in Bynum, W. and Porter, R. (eds.), *William Hunter and the Eighteenth-Century Medical World*, pp. 153–76, Cambridge University Press, Cambridge.
20. Littlejohn, H. D.: 1897, Lecture Notes, 18 May 1897, DK.3.41, Edinburgh University Library, Edinburgh.
21. Littlejohn, Harvey: 1922/4, MS Notes of Lectures, Gen 1992/4, Lecture 36, Edinburgh University Library, Edinburgh.
22. Mann, J. D.: 1893, *Forensic Medicine and Toxicology*, Charles Griffin & Co., London.
23. Ogston, F.: 1878, *Lectures on Medical Jurisprudence*, J. &. A. Churchill, London.
24. Paris, F. J. and Fonblanque, J. S.: 1823, *Medical Jurisprudence*, W. Phillips, London.
25. Percival, T.: 1803, "Medical Ethics; or, a Code of Institutes and Precepts Adapted to the Professional Conduct of Physicians and Surgeons," in *Works, Literary, Moral and Medical*, Vol. 2, pp. 355–456, J. Johnson, London.
26. Tomaselli, S. and Porter, R. (eds.): 1986, *Rape*, Blackwell, Oxford.
27. Public Record Office, file LCO2/624, London.
28. Risse, G. B.: 1986, *Hospital Life in Enlightenment Scotland*, Cambridge University Press, Cambridge.
29. Roughead, W. (ed.): 1906, *Trial of Dr. Pritchard*, William Hodge & Co., Glasgow/Edinburgh.
30. Scottish Record Office: file AD 56/17 (22 Sept. 1853), Edinburgh.
31. Smith, J. G.: 1821, *Principles of Forensic Medicine Systematically Arranged, and Applied to British Practice*, Underwood, London.
32. Smith, J. G.: 1825, *An Analysis of Medical Evidence Comprising Directions for Practitioners, in the View of Becoming Witnesses in Courts of Justice*, Underwood, London.
33. Taylor, A. S.: 1844, *A Manual of Medical Jurisprudence*, John Churchill, London.
34. Taylor, A. S.: 1984, *Taylor's Principles and Practice of Medical Jurisprudence*, 13th ed., K. Mant (ed.), Churchill Livingstone, Edinburgh.
35. Waddington, I.: 1975, "The Development of Medical Ethics: A Sociological Analysis," *Medical History* 19(1), 36–51.
36. Waddington, I.: 1984, *The Medical Profession in the Industrial Revolution*, Gill & Macmillan, Dublin.
37. McLaren, A.: 1993, "Priviliged communications: medical confidentiality in late Victorian Britain," *Med. Hist.*, 37(1), 129–147.

CHAPTER 9

PETER BARTRIP

SECRET REMEDIES, MEDICAL ETHICS, AND THE FINANCES OF THE *BRITISH MEDICAL JOURNAL*

This essay explores three themes relating to the use of patent or proprietary medicines, the so-called "secret remedies," in the Victorian and Edwardian periods. First, quackery in nineteenth-century Britain and in particular the attitudes of the Provincial Medical and Surgical Association (British Medical Association, or "BMA", from 1856) towards it. Second, the economics of the *British Medical Journal* (*BMJ*) and its relationship with its parent body, the BMA: How was the *Journal* funded? Was it profitable? To what extent was the Association reliant upon *BMJ* advertising revenue? Third, the *BMJ*'s well-known campaign against the trade in secret remedies. These themes will then be drawn together by examining how they related to *Journal* policy on the acceptance of advertising from pharmaceutical and other companies. The question to be asked here is: was this policy ethical?

QUACKERY AND THE MEDICAL PROFESSION

The popular notion of a quack, it has been suggested, is "of an ignorant and unscrupulous pretender, often itinerant, who preyed on a credulous public for profit; a confidence trickster" ([20], p. 14). Modern historiography, however, is more circumspect, tending to eschew the word "quack" in favour of more neutral terms such as irregular, fringe, or unorthodox practitioner. When referring to the early nineteenth or previous centuries there are sound reasons for avoiding pejoratives, for with medical training and qualification highly variable, it was often far from clear precisely who was the quack. A university education was an unreliable guide when, in 1818, the Professor of Anatomy at Trinity College Dublin could comment on "the deadness of things medical" in

R. Baker (ed.), The Codification of Medical Morality, 191–204.
© 1995 *Kluwer Academic Publishers. Printed in the Netherlands.*

Oxford University ([21], p. 122). Neither did inclusion in the *London and Provincial Medical Directory* necessarily count for much. "I think I could select a name or two," wrote a correspondent to the *Provincial Medical and Surgical Journal*, "for which their owners (as men of letters), can make no higher claim than appending to their patronymics now in that Directory, the letters M.S.O.F., that is to say, Member of the Society of Odd-Fellows" ([31], 5 May 1847, p. 252). As for the medical corporations, which were supposed to regulate the profession, these were, in reality, unable to prevent the unqualified from practising or even to warrant the skill and probity of their members. Thomas Wakley, it should be remembered, built the reputation of the *Lancet* by exposing quackery and incompetence in high, as well as in low places.

The distinction between regular and irregular practitioners was further blurred by several factors relating to the use and abuse of medicines. First, the "heroic" prescribing practices of many of the regular profession – large doses, frequently administered – which, it is now recognised, may often have done more harm than the ministrations of the unqualified ([10], p. 227; [26], Chapter 7). Second, the government stamp on proprietary medicines (introduced in 1783 as a means of raising revenue) appeared to confer a spurious respectability on them, amounting, not only in the eyes of their proprietors, to an official seal of approval ([34], p. 14; [14], 11 March 1857, p. 155; [17], 4 Feb. 1857, pp. 171–2). Third, and perhaps most important, the active involvement of the orthodox profession with quackery. As Porter has noted, with reference to the 1830s, "top practitioners were up to the chin in quackish practices," including endorsement of patent medicines ([29], p. 226). Hence, the ethical rhetoric was not always in step with marketplace reality.

John Gregory's eighteenth-century, *Lectures on the Duties and Qualifications of a Physician* provided copious guidance in matters of medical ethics, including on professional behavior with regard to nostrums and secret remedies. Gregory argued that one of "the duties the physician owes to his patients" is that of "allowing them every indulgence consistent with their safety." Such indulgence included listening to suggested remedies, and not only because "every man has a right to speak where his health or his life is concerned." It was also possible that the "proposal may be a good one" and therefore worth trying. Gregory went on to criticize faculty who rejected remedies proposed by patients or their friends "from a pretended regard to the dignity of the profession, but in reality from mean and selfish views." If the patient was determined to try a remedy, the physician could convey his disapproval, but, Gregory thought that the physician had no right to complain if his advice was rejected ([15], pp. 32–3).

Elsewhere, in the same lecture, Gregory showed a degree of sympathy with physicians who kept secret remedies or nostrums. He acknowledged that "the bulk of mankind" has more respect for something imbued with an aura of mystery and complexity. The implication was that this respect could contribute to the healing process. Hence, "when a nostrum is once divulged, its wonderful qualities immediately vanish." Only after a balanced discussion did Gregory come out against nostrums, and then, somewhat grudgingly: "I am persuaded," he said, "that nostrums, on the whole, do more harm than good." While allowing that continental physicians of honour and reputation kept nostrums, he concluded that "still the practice has an interested and illiberal appearance." Gregory even showed reluctance to criticize vendors of quack medicines: "A vender of quack medicine does not tell more lies about its extraordinary virtues, than many people do who have no interest in the matter; even men of sense and probity." To summarize, Gregory was opposed to secret remedies, but willing to admit not only that there was something to be said in their favor, but that physicians should accept the right of their patients to experiment. His, therefore, was a relaxed and reasonable attitude ([15], pp. 59–61).

Gregory's treatment of quack medicines was picked up and modified by Thomas Percival, whose highly influential *Medical Ethics* was first published in 1803. In Chapter Two, Article XXI, Percival stated that:

> The use of *quack medicines* should be discouraged by the faculty, as disgraceful to the profession, injurious to health, and often destructive even of life. Patients, however, under lingering disorders, are sometimes obstinately bent on having recourse to such as they see advertised, or hear recommended, with a boldness and confidence, that no intelligent physician dares to adopt with respect to the means that he prescribes. In these cases, some indulgence seems to be required to a credulity that is insurmountable: And the patient should neither incur the displeasure of the physician nor be entirely deserted by him. He may be apprized of the fallacy of his expectations, whilst assured, at the same time that diligent attention should be paid to the process of the experiment he is so unadvisedly making on himself, and the consequent mischiefs, if any, obviated as timely as possible. Certain active preparations, the nature, composition and effects of which are well known, ought not to be proscribed as quack medicines.

In this section, which is recognizably "Gregorian" in origin, Percival repeats, clarifies and strengthens his predecessor's position. He emphasizes the physician's responsibility to discourage the use of quack medicines but acknowledges that patients intent on taking them should be shown indulgence. In the last sentence of the Article, Percival makes a further concession to toleration, recognizing the acceptability to the profession of "over the counter" drugs provided that they were not secret remedies.

It was in Article XXII, dealing with the conduct of physicians themselves, rather than what they should be prepared to tolerate with respect to patients' wishes, that Percival adopted a much tougher line than Gregory:

> No physician or surgeon should dispense a secret *nostrum*, whether it be his invention, or exclusive property. For if it be of real efficacy, the concealment of it is inconsistent with beneficence and professional liberality. And if mystery alone give it value and importance, such craft implies either disgraceful ignorance, or fraudulent avarice ([28], pp. 44–45; [38]).

This pronouncement, definite and unequivocal as it was, was echoed by later authors and codifiers, including the American Medical Association in 1847, and by some from within the PMSA/BMA. Take the Manchester Medical Ethical Association, which was formed in 1847 with a committee dominated by PMSA members. Of its eight rules, breach of which meant expulsion or refusal of membership, three referred to patent medicines:

1. No member shall be the proprietor of, or in any way derive advantage from, the sale of any patent or proprietary medicine.
2. No member shall give testimonials in favour of any patent or proprietary medicine, or in any way recommend their public use.
3. No member, who may keep an open shop, shall sell patent medicines, perfumery, or other articles than pharmaceutical drugs and preparations ([31], 11 Aug. 1847, p. 448; 17 Nov. 1847, pp. 639–41).

These rules spelled out Percival's Article XXII while remaining silent about the indulgence discussed in Article XXI.

Clearly some of these "post-Percivaleans," went somewhat further than their mentor in their strictures against professional involvement with proprietary remedies. Percival had urged "some indulgence" where patients "under lingering disorders" insisted on having a medicine which they had seen advertised or to which they had been recommended. Furthermore: "Certain active preparations, the nature, composition, and effects of which are well known, ought not to be proscribed as quack medicines" ([28], pp. 44–5). However, half a century later Jonas Malden, a leading light in the PMSA and a close friend of its founder, Charles Hastings, said: "There are no secrets in true science, – no private purposes to serve by it. . . . *A nostrum should not be so much as named among us*" ([22], pp. 222–3, *emphasis added*). In 1878 Jukes Styrap, a founder of the Shropshire Branch of the BMA (see Appendix C) wrote in his *Code of Medical Ethics*, that it was:

derogatory to professional character...for a practitioner to hold a patent for any proprietary medicine or surgical instrument; or to dispense a secret *nostrum*, whether it be the composition, or exclusive property of himself, or of others: for, if such *nostrum* be really efficacious, any concealment in regard to it is inconsistent with true beneficence and professional liberality; and if mystery alone impart value and importance to it, such craft is fraudulent ([34], pp. 27–8).

Styrap's code, like the AMA's, owed more than a little to Percival whose pioneering work retained its relevance even seventy-five years after its first publication ([34], p. 3). But of more interest than Styrap's "duplications" are his omissions from and additions to the original text. Hence, when discussing quack nostrums, the one section of Percival which Styrap ignored was that which dealt with the circumstances in which tolerance was advisable. This omission suggests a hardening of professional attitudes, as does Styrap's inclusion of an additional precept: "It is also extremely reprehensible for a practitioner to attest the efficacy of patent or secret medicines, *or, in any way, to promote their use* . . ." ([34], pp. 27–8, *emphasis added*). This was more redolent of Malden than of Percival's subtler prescriptions. Indeed, drawing a line from Gregory to Styrap, it is clear that in slightly less than 100 years the ethical position of the British profession with regard to quack medicines showed a clear hardening of attitude.

Of course, the enunciation of principles is not necessarily synonymous with their observance. After all, the professional ethics discussed here lacked the authority and sanction of the State. In fact, as Charles Cowan of the Provincial Medical and Surgical Association made clear when he read a report on Empiricism to the Association's annual meeting in 1840, many regular practitioners were involved in the quack medicine trade. Not only did they endorse certain proprietary remedies by allowing their names to be used in advertising, but, Cowan alleged, "instances exist where medical men in the metropolis, of supposed respectability of conduct, are secretly the owners of establishments for the sale of quack medicines" ([35], IX, 1841, pp. 44–5; [7]). This prompted Thomas Jeffreys to enquire: "was it not well-known and acknowledged that a great proportion of the medicines daily prescribed by the faculty were quackish in their origin?" A dozen years later the editor of the *Association Medical Journal* lamented that "cheating advertisements of pills, ointments and nasty books" were not regarded by the profession "with the abhorrence which they deserve" ([1], 21 Jan. 1852, p. 55). Many similar examples could be given.

So the line between orthodoxy and unorthdoxy could be vague, not least in respect of drug prescription and endorsement, and the position of some regulars did not bear close examination ([23], p. 245). Yet many of those who counted themselves qualified regulars were keen to firm up the distinction by banning unqualified practice. Quackery, as Professor James Macartney said as early as 1838, "ought to be visited with the most severe legal punishment, or at the least to be made equal with the crime of obtaining money under false pretences" ([35] VII, 1839, p. 34). Such opponents of quackery tended to justify their views in terms of a determination to protect a gullible public which was being defrauded by its purchase of useless and sometimes dangerous preparations, but there can be little doubt that they also wished to monopolize treatment, advance their status and respectability, improve the market for their services or, in the sociological jargon, achieve "professional closure." So when, in the early nineteenth century, the regular profession began to complain long and loud about the quacks, it reveals, as Roy Porter has argued, more about "the politicisation of medicine than of the fortunes of quackery itself" ([29], p. 222).

The Provincial Medical and Surgical Association was founded in Worcester in 1832. At first it took little interest in medical politics, being more concerned to foster medical science in provincial England. But from the late 1830s, as the medical reform question, began to "warm up," it paid increasing attention to political issues. One of its goals was to secure an Act of Parliament which would suppress quackery, for quacks, far more than the medical corporations against which Wakley railed, offered the main threat to its members, most of whom lived in England, outside London. The *Provincial Medical and Surgical Journal* (*BMJ* from 1857) became the Association's main weapon for beating the quacks and advocating reform. Before considering the *Journal*'s role in the secret remedies campaign, we need to consider the manner in which it was financed, for this had important implications for its assault on quackery.

THE FINANCES OF THE *BMJ*

For various reasons the 19th century saw huge growth in the number of medical journals including, from 1823, weeklies which dispensed a varied diet of news, opinion, scholarly articles and so forth ([2], pp. 6–12, [12], [18]). The most successful weeklies were the *Lancet*, the *Medical Times and Gazette*, the *Medical Press and Circular* and the *BMJ*. Although their formats were broadly similar, there was one important respect in which the *BMJ* differed from its competitors. While they were all commercial speculations, as it was to begin with, it soon became the "organ of the Association." As such it has gone, since the early 1840s, to all Association members as a benefit of membership. Copies have always been available for purchase by non-members, but non-member sales have never comprised more than a small proportion of the print run. So while sales have provided a major part of its rivals' income, this has never been the case for the *BMJ*. In 1896, for example, sales revenue amounted to just over £1700, but this covered only about 7 per cent of editorial and production costs. The deficit was made good in two ways: by advertising revenue and subvention from BMA funds. In 1896 advertising income was about £15,500. In order to balance the books the BMA had to hand over some £7000 of subscription income to the Journal Department. This year was no freak. Until comparatively recently the *BMJ* has been funded (in descending order of importance) by advertising, subvention from BMA subscriptions and sales.

In contrast with its American counterpart, the American Medical Association, no part of BMA subscriptions have ever been earmarked in advance for its *Journal*. Practice has always been for Council (the BMA's governing body) to take surpluses and make good deficits. In the 19th century there never were any surpluses. This had several implications. It diminished the *BMJ*'s independence, helped create the impression that it was a loss-making operation and thereby to foster criticism of it by BMA members, and encouraged those who ran the *Journal* to chase advertising.

With the appointment of Ernest Hart as editor in 1866 (he served between 1867 and 1869 and again between 1870 and 1898) the need to maximize advertising revenue increased, for Hart made it clear when he was appointed that he intended to spend his way to success. With longstanding complaints from some BMA members about the amount of subscription income being pumped into the *Journal*, and with limited opportunity for increased sales, his only means of generating significant extra income was through advertising ([4], [9], 1 July 1871, p. 20). Under Hart's editorship advertising rates changed little, but advertising revenue grew dramatically owing to the increased space devoted to advertisements. By 1889 individual issues of the *BMJ* were carrying 62 pages of advertising as against 56 of editorial matter. Ten years later the ratio had become 80:64 in favour of advertising. Without advertising revenue BMA subscription would have had to be far higher or the *BMJ* would have been a very different publication. How higher subscriptions or a more limited *Journal* would have reduced membership is, of course, a moot point.

So what was advertised in the *BMJ*? This is not easily established because few volumes of the *Journal* were bound and preserved with their advertisement pages. Rather, these were removed and destroyed, partly, no doubt, owing to problems of binding and storing such bulky material, but also because advertisements were felt to lack enduring interest. To facilitate their easy removal they were placed at either end of the *Journal* rather than being interspersed with editorial matter. Surviving advertising shows that the *BMJ* carried advertisements for almost anything that doctors and their families, friends and patients might buy, including musical instruments, food, drink, cigarettes, clothing and holidays, as well as more strictly "medical" items. Medical appointments, books, equipment and journals were always prominent. Specific products changed over time and reflected social and other trends. Homes for inebriates and others in need of long-term care were to the fore by the eighties, as were bicycles. Motor manufacturers were purchasing considerable space by the end of the century. Even without undertaking a systematic analysis of advertising content, it is clear that "chemical and medicinal preparations" loomed large and provided a substantial proportion of the revenue which was so vital to both the *Journal* and the BMA.

THE *BMJ's* CAMPAIGN AGAINST SECRET REMEDIES

With the passage of the Medical Act of 1858, the BMA's concern about the quack medicine trade seems to have diminished somewhat, thereby endorsing Porter's point that earlier complaints were related to the movement for medical reform. But from around 1880 the Association, mainly through the *BMJ* and its Parliamentary Bills Committee (both of which were dominated by Ernest Hart) again began to campaign strongly for controls on proprietary articles ([19], pp. 124–5). This was probably a response to the tremendous surge in sales: the number of licensed vendors increase by around 55 per cent, and receipts from the government stamp by close to 90 per cent, during the 1870s. By the 1880s the patent medicine trade was thought to support as many as 1,000 owners and 19,000 employees; total sales amounted to some £1.5m per year, and the stamp duty yielded government revenue in excess of £200,000 per annum. This boom is usually explained in terms of an improved working class standard of living (more disposable income) combined with the introduction of legal restrictions on the sale of opium preparations, and a reduction in medicine license duty ([3], p. 125; [13], pp. 22–3, 203–4; [33], pp. 343–5; [36], pp. 40–6, 61–8). It was fueled by the lay press which attracted acres of highly remunerative proprietary medicine advertisements. Indeed, it is plausible that the late Victorian newspaper boom was largely financed by the ready availability of such advertising.

The *BMJ*'s campaign employed a variety of strategies, including support for legislation restricting sales to registered pharmacists and obliging manufacturers to reveal ingredients, criminal prosecution of those in breach of the law, and the persuasion of newspaper proprietors to cease "prostituting their advertisement columns...to the service of a dangerous and often wicked class of imposters" ([9], I, 27 Jan. 1894, p. 208). In 1904, Hart's successor, Dawson Williams, commissioned a distinguished pharmacist, Edward Harrison, to analyze a wide range of proprietary medicines with a view to identifying contents ([24], pp. 254–5). Williams published Harrison's findings in the *BMJ* along with details of the (usually high) price of particular products contrasted with the (usually minute) value of their ingredients. The series ran on an occasional basis till the end of 1908, by which time dozens of products claiming to cure almost every conceivable ailment had been exposed as valueless. In 1909, many of the articles were reproduced in book form as *Secret Remedies. What they Cost and What they Contain* [32]. It was tremendously successful, selling around 62,000 copies in less than two years. A second, less successful, volume, entitled *More Secret Remedies*, was published in 1912. In addition to analyses of quack medicines, this contained sections on "unqualified practice through the post," the advertising of medicines and the "expert" behind them [25].

ADVERTISING ETHICS

The campaign against secret remedies has been seen not only as brilliantly innovative medical journalism, but also as one of the *BMJ*'s (and the BMA's) shining achievements ([37], pp. 91–105; [19], pp. 273–7). There was, however, little truly original either in the medical press attacking nostrum mongers and proclaiming the worthlessness of their goods, or in it criticizing the part played in the trade by lay newspapers. The *Lancet*, for example, had revealed the ingredients of quack medicines as long ago as the 1820s. In the 1850s, John Rose Cormack, editor of the *Association Medical Journal* (soon to be renamed the *BMJ*) had persuaded the editor of the *Critic*, a London literary journal, to cease taking advertising for Holloway's pills. Cormack even chided the *Lancet*, normally an arch opponent of all quackeries, for allowing the insertion of a quack advertisement: "If you denounce quackery in vehement leaders, do not sell it a shelter in your advertising columns." In 1865 the *Dublin Medical Press* listed eighteen British and Irish newspapers which had been persuaded to forego quack advertising ([1], 21 Jan. 1852, p. 55; [14], 8 March 1865, pp. 235–6; [17], I, 1823,p. 62; 15 April 1837, pp. 130–1; [29], p. 223). Also, if we look at what the *BMJ* campaign achieved, it is hard to point to anything very specific. Certainly, the proprietary medicine trade was in no way curtailed; neither can the *BMJ* be directly credited with the appointment of the 1912 Select Committee on Patent Medicines which, in any case, had little immediate impact owing to the outbreak of the First World War. In fact, there are grounds for arguing that the main significance of the secret remedies campaign is in underlining the hypocrisy and dubious ethics of the Association and its *Journal*.

After the *BMJ* published its first paper on secret remedies there was an eighteen-month delay before further exposees were published. While this gap owed something to Edward Harrison's illness, another factor was receipt of a letter from the editor of the *Journal of the American Medical Association*, George Simmons, which raised doubts as to whether the *BMJ*'s own advertising columns were themselves free of secret remedies. Simmons, who explained that the AMA was conducting its own campaign against secret remedies, listed six such products which were being advertised in the *BMJ*, four of which had been denied access to *JAMA*. His main objection to these was that they were being advertised in the lay press and were available to the public "over the counter," though he was also concerned that their composition was not disclosed and hence that they might be prescribed by doctors ignorant of their contents, simply on the claims of their manufacturers ([5], XIV, 10 Jan. 1906, p. 214; [11], pp. 67–92, 107–31).

Responsibility for *BMJ* advertising rested with its business manager Guy Elliston, who was also the BMA's general secretary. He claimed that he rejected copy for "certain articles largely advertised in the lay press" and also for medicines "at least those intended for internal administration, the ingredients of which are not stated in the advertisement," liaising with the editor on doubtful cases. The Journal and Finance Committee, having investigated, set up an advisory committee but decided that the manager should continue to accept advertisements at his discretion provided he was informed of the ingredients of all preparations advertised. In practice, the *Journal* continued to carry advertisements for medicines the composition of which was not disclosed.

In 1909 complaints about advertisements containing misleading or exaggerated statements were received from the BMA's Birmingham branch, but these were to no more effect than Simmons's had been. Many dubious products continued to be advertised in the *BMJ*. Hence, in 1910 the National British Women's Temperance Association complained about the *Journal* carrying advertisements for "Wincarnis," and other so-called medicated wines, which it termed "one of the great class of quack secret remedies which are ruining the Medical Profession and the public." The NBWTA accepted that the advertisements were "a lucrative source of income," and also that their appearance in *BMJ* advertising columns did "not necessarily imply" approval, nevertheless it called for the *Journal* to be closed to such products. In this instance, urged on by the BMA's Annual Representative Meeting, subsequent advertising for Wincarnis and Hall's Wine, a product which contained cocaine, was rejected ([5], XVII, 26 Nov. 1909, pp. 1240–1, 1253; XVIII, 22 June 1910, p. 1300; 19 Oct. 1910, p. 1677; 21 Dec. 1910, p. 1855; [6], 26 Oct. 1910, pp. 1539–41). Sanatogen, diabetic whisky, other alcoholic drinks and a variety of secret remedies, on the other hand, continued to appear.

In 1912, no longer content with confidential remonstration, *JAMA* criticized the advertising practices of the *BMJ* (and other UK medical periodicals, including the *Lancet*) in its editorial columns. The BMA, it said:

> has...neglected to clean its own skirts. . . . No attempt has been made . . . to purge the British medical profession of the innumerable fraudulent proprietary remedies with which it is afflicted . . . nostrums . . . hold high revel in the advertising pages of high-class medical journals in Great Britain. . . . Wherein is it any worse for the public to buy medicinal preparations about which it knows nothing, than it is for medical men to prescribe medicinal preparations about whose composition they know nothing? ([16], 3 Feb. 1912, p. 349).

JAMA predicted that the proprietary medicine interest would be able to resist demands for regulation by pointing to the unsatisfactory practices of the medical profession. The recently appointed select committee would, therefore, achieve little.

Sure enough, when Alfred Cox, the BMA's newly appointed medical secretary appeared as a witness before the select committee he was subjected to what he subsequently called "a very searching examination." The BMA's written evidence had called attention to "the extent to which financial considerations apparently outweigh all considerations of honourable responsibility on the part of many of the newspapers in this country." But when confronted with the 27 January 1912 issue of the *BMJ*, Cox had to agree that, of its first 34 pages, about 16 were filled with proprietary medicine advertisements. As he later admitted in private, "it was evident to me that some of these advertisements were so worded as to make the critical attitude of the Association to other Journals rather difficult to maintain" ([8], 10 Jan. 1913, pp. 2–4; [27], pp. 127–30).

After his grilling Cox reported his misgivings to the BMA's Journal and Finance Committee, but in the discussion which followed Elliston pointed out that to reject all the advertisements criticized by the Select Committee would cost the Association around £ 1400 per year. Little was agreed other than to reject future advertising for Sanatogen. The *Journal* continued to accept copy for Angiers Emulsion which was one of the products criticized by both *JAMA* and by the Select Committee. Thereafter, in every decade up to and including the 1960s (the last for which I have examined BMA files) serious criticisms were raised either by academics, BMA members or other medical associations about one or more of the food or drug preparations advertised in the *BMJ*.

The simple conclusion to all this is that the *BMJ*, in flouting its self-proclaimed canons of morality, acted in a hypocritical manner. Moreover, in profiting from the advertising of secret remedies it was in clear breach of a well-established ethical tradition which, as we have seen, counseled the profession to avoid all contact with secret medicines. Intent on attacking others for making money out of proprietary articles, it behaved very differently when it was itself in a position to profit from the trade. The secret remedies campaign may therefore be dismissed merely as an example of a powerful interest group advocating "ethical standards" which it failed to meet itself. But is this a high-minded and unrealistic assessment? After all the *Journal* did exclude many advertisements involving it in considerable loss of revenue. In addition, it would appear that its rival publications often carried similar advertising. The *BMJ* was not therefore uniquely culpable. Even *JAMA* was, self-confessedly, carrying advertisements for proprietaries in 1905 ([11], p. 74). As for the accusations of medical journals (such as *JAMA*) against their rivals, these must be seen in the context of the competition for readers; mud-slinging was a part of the circulation war. Not many years since, *JAMA* had had a specific grievance against the *BMJ* when Ernest Hart, having been consulted about how *JAMA* might be improved, contacted a large number of American practitioners urging them to subscribe to the *BMJ* instead of its U.S. counterpart ([37], p. 135). So *JAMA*'s criticisms need to be critically assessed.

To be sure, the *BMJ* was caught in a cleft stick; obliged to rely on advertising to survive, it was also expected to exclude all manner of advertisements lest they offend members or contradict BMA policy. As the organ of a mass membership medical society the *BMJ* faced a harder task in accepting and rejecting advertising copy than many other parts of the press, including, perhaps, other medical periodicals, for it was a journal in whose direction, it has been said, every BMA member claimed a voice ([30], LX, 1898, pp. 117–8). Certainly the lay press was much less likely ever to be accused of "in effect giv[ing] its blessing" to a product simply by carrying an advertisement for it. Yet this was what Professor Yudkin of London University's Department of Nutrition argued when, in 1957, he criticized the *BMJ* for carrying an advertisement for a slimming aid – "Larson's Swedish Milk Diet" – which he considered to be valueless.

The *BMJ* has discovered that advertising can be an area fraught with dilemmas on numerous occasions, including contraceptives in the 1930s, which outraged Roman Catholic members of the BMA, and cigarettes, which were dropped in the late fifties – though only some years after Richard Doll and Austin Bradford Hill had demonstrated, in the *BMJ*, a link between tobacco smoking and lung cancer. Even an advertisement for *Soviet Weekly* drew a letter of protest in 1954. Even so, once all the allowances have been made, the secret remedies campaign and its aftermath still reveal a *Journal* that advocated a standard of ethics which it regularly failed to achieve itself.

Finally, the episode provides a pointer to how medical historians should approach questions of ethics. They should avoid the trap of thinking of these simply in terms of written codes of conduct compiled by committees or individuals. Ethics might propose ideals but they are also concerned with regulating real relationships, whether between practitioners or between doctors and patients. As such, they need to be assessed empirically within an "action framework." It should not be thought that ethical codes necessarily correspond with prevailing practice. Indeed, the history of the *BMJ*'s secret remedies campaign, viewed in the context of the *Journal*'s own involvement with the proprietary medicine trade, draws attention to one gap which existed between rhetoric and reality. Whether similar gaps existed elsewhere in the profession is a question historians would be well advised to consider.

BIBLIOGRAPHY

1. *Association Medical Journal.*
2. Bartrip, P.: 1990, *Mirror of Medicine. A History of The British Medical Journal*, Clarendon Press, Oxford.
3. Berridge, V. and Edwards, G.: 1987, *Opium and the People: Opiate Use in Nineteenth Century England*, Yale University Press, New Haven and London.
4. British Medical Association: 1869, Mss. Committee of Council Minute Book 1868–75. Copy of Letter from Ernest Hart dated 20 June 1868, pp. 8–13.
5. British Medical Association mss. Minutes of Committees and Sub-Committees, Journal and Finance Committee.
6. British Medical Association mss. Minutes of Council.
7. British Medical Association: 1841, Mss. Provincial Medical and Surgical Association Minutes, 1834–1847. An Account of the Proceedings of the Ninth Anniversary Meeting.
8. BMA mss. Signed Committee Minutes, Journal Committee.
9. *British Medical Journal.*
10. Brown, P. S.: 1987, "Social Context and Medical Theory in the Demarcation of Nineteenth Century Boundaries," in Bynum, W. and Porter, R. (eds.), *Medical Fringe and Medical Orthodoxy*, pp. 216–33, Croom Helm, London.
11. Burrow, J. G.: 1963, *AMA Voice of American Medicine*, Johns Hopkins Press, Baltimore.
12. Bynum, W.F., Lock, S. P., and Porter, R. (eds.): 1992, *Medical Journals and Medical Knowledge*, Routledge, London.

13. Chapman, S.: 1974, *Jesse Boot of Boots the Chemist A Study in Business History*, Hodder and Stoughton, London.
14. *Dublin Medical Press.*
15. Gregory, J.: 1772, *Lectures on the Duties and Qualifications of a Physician*, Strahan and Cadell, London.
16. *Journal of the American Medical Association.*
17. *Lancet.*
18. Le Fanu, W. R.: 1984, *British Periodicals of Medicine*, rev. ed., Wellcome Unit for the History of Medicine, Oxford.
19. Little, E. M.: 1932, *History of the British Medical Association*, British Medical Association, London.
20. Loudon, I.: 1986, *Medical Care and the General Practitioner*, Clarendon Press, Oxford.
21. Macalister, A.: 1900, *James Macartney. A Memoir*, Hodder and Stoughton, London.
22. Malden, J.: 1851, "On Empiricism," *Transactions of the Provincial Medical and Surgical Association* XVIII, pp. 219–228.
23. Marland, H.: 1987, *Medicine and Society in Wakefield and Huddersfield, 1780–1870*, Cambridge University Press, Cambridge.
24. Matthews, L. G.: 1962, *History of Pharmacy in Britain*, Livingstone, Edinburgh/London.
25. *More Secret Remedies* (1912), BMA, London.
26. Nicholls, P.: 1988, *Homeopathy and the Medical Profession*, Croom Helm, London.
27. *Parliamentary Papers*: 1914, IX, Select Committee on Patent Medicines, Evidence.
28. Percival, T.: 1803, *Medical Ethics*, Johnson and Bickerstaff, Manchester.
29. Porter, R.: 1989, *Health for Sale. Quackery in England, 1660–1850*, Manchester University Press, Manchester.
30. *Practitioner.*
31. *Provincial Medical and Surgical Journal.*
32. *Secret Remedies. What they Cost and What they Contain* (1909), BMA, London.
33. Smith, F. B.: 1979, *The People's Health, 1830–1910*, Croom Helm, London.
34. Styrap, J.: 1878, *A Code of Medical Ethics*, Churchill, London; this volume, pp. 206–239.
35. *Transactions of the Provincial Medical and Surgical Association.*
36. Turner, E. S.: 1952, *The Shocking History of Advertising*, Michael Joseph, London.
37. Vaughan, P.: 1959, *Doctors' Commons. A Short History of the British Medical Association*, Heinemann, London.
38. Waddington, I.: 1975, "The Development of Medical Ethics – A Sociological Analysis", *Medical History* 19, 36–51.

CHAPTER 10

RUSSELL G. SMITH

LEGAL PRECEDENT AND MEDICAL ETHICS: SOME PROBLEMS ENCOUNTERED BY THE GENERAL MEDICAL COUNCIL IN RELYING UPON PRECEDENT WHEN DECLARING ACCEPTABLE STANDARDS OF PROFESSIONAL CONDUCT[1]

INTRODUCTION

In 1858, legislation was enacted in Britain which created a statutory body entitled the General Council of Medical Education and Registration of the United Kingdom [13]. This title, which was subsequently described by Viscount Addison in the House of Lords as "a cumbersome name" the initials of which "have often befogged many of us in the past" ([9], p. 913) and which was later abbreviated to General Medical Council or "G.M.C." [14], contains reference to the two principal functions of the Council: the determination of whose name should be permitted to be placed upon the Medical Register; and the determination of whose name should be removed from the Register. The G.M.C. performed the first function by setting and monitoring educational standards for doctors; it performed the second by holding judicial, disciplinary inquiries. In exercising these functions, the Council, through its constituent bodies and Committees, indirectly became involved in the declaration of acceptable standards of professional conduct and medical ethics.

At present, the Council has a legislative power to provide advice to members of the profession on standards of professional conduct and medical ethics and this function is undertaken principally by the Council's Committee on Standards of Professional Conduct and on Medical Ethics, the "Standards Committee" ([15], s. 35). This Committee is directly involved in the preparation and revision of the guidance given to all registered medical practitioners in the form of a small, forty-two page, blue-covered booklet entitled "Professional Conduct and Discipline: Fitness to Practise" [7].

R. Baker (ed.), The Codification of Medical Morality, 205–218.
© 1995 *Kluwer Academic Publishers. Printed in the Netherlands.*

In the 1850s, however, when the G.M.C. was first established, no such formal guidelines of good professional conduct existed but, rather, they developed out of the decisions handed down by the Council in disciplinary cases decided over the years.

This essay will examine how this process of declaring acceptable standards of professional conduct by the use of decided cases developed, and will raise a number of problems which, arguably, it created in the nineteenth century and which continue to cause concern today.

DISCIPLINARY JURISDICTION AND PROCEDURE IN THE NINETEENTH CENTURY

The G.M.C.'s power to hold disciplinary inquiries into the professional conduct of practitioners originally derives from section XXIX, of the Medical Act of 1858:

> XXIX. If any registered Medical Practitioner shall be convicted in England or Ireland of any Felony or Misdemeanor, or in Scotland of any Crime or Offense, or shall after due inquiry be judged by the General Council to have been guilty of infamous conduct in any professional Respect, the General Council may, if they see fit, direct the Registrar to erase the Name of such Medical Practitioner from the Register.

The Medical Act 1858 came into operation on 1 October 1858; as early as June 1860, the Council was called upon to institute an inquiry under section XXIX in the case of Richard Organ ([6], I, August, 11, 1859, p. 69; June 14, 1860, p. 81; June 15, 1860, pp. 83–94; June 18, 1860, pp. 94–7; June 19, 1860, pp. 103–4). No specific procedure had been laid down by Parliament, save for an obligation to make "due inquiry" and the Council accordingly adopted a quasi-criminal procedure following the advice of its lawyers. This first attempt at a disciplinary inquiry resulted in a practitioner's name being erased from the Register; however, the practitioner successfully applied to the High Court to have his name restored on the Register, on the grounds that he had been denied an opportunity of being heard [21]. Following this unfortunate start, the Council appointed a Committee to prepare regulations for the hearing of proceedings under sections XXVI and XXIX Medical Act of 1858 ([6], II, July 1, 1861, p. 20). These were adopted on 3 July 1861, generally employing a judicial, adversarial, quasi-criminal model of procedure ([6], II, July 3, 1861, pp. 42–3).

Initially, the Council was composed of twenty-four members who were all able to be involved in the hearing of disciplinary cases, and usually most members of the Council heard any given case [4]. This contrasts with the position which obtained after 1950 ([14], s. 14) when smaller Committees were given jurisdiction to deal with disciplinary cases. All the members of the G.M.C. were medically qualified until 1926. In that year the first lay member was appointed to the Council ([6], November 27, 1956, XCIII, p. 32), following a celebrated disciplinary case ([6], XLVIII, May 24, 1911, pp. 52–4) and, in part, owing to a concerted campaign by George Bernard Shaw to have a majority of lay members on the G.M.C. ([19], pp. 47–8). The function of lay members on the Council is essentially to act as disinterested members of the public who are able to represent the consumer's interest in debates. However, when hearing disciplinary cases, a certain tension arises between their non-medical independence and their function as adjudicators of matters which sometimes require medical training and expertise.

Owing to the fact that G.M.C. members who sit on disciplinary cases usually do not possess legal qualifications, a good deal of reliance is placed upon the lawyers who appear on behalf of the Council or complainant and the accused practitioner to ensure that cases are conducted in accordance with substantive and formal requirements of law. In the 1880s, it became the practice to have Council's solicitor present to conduct the case against the accused practitioner, and also to have a barrister present, acting as a judicial assessor giving advice on questions of law and ruling on the admissibility and weight of testimony. The legal assessor, however, has never been formally involved in voting during the adjudication of cases. The first legal assessor to be present for a hearing on 26 April 1881, was the writer Charles Dickens' son, Henry Fielding Dickens ([6], April 26, 1881, XVIII, p. 68).

DISCIPLINARY CASES 1858 TO 1883

Between 1858 and 1883, registered medical practitioners were given no written guidance by the G.M.C. as to how they should regulate their professional conduct so as to avoid disciplinary action. By reading reports of the early disciplinary cases in the medical and lay press, they could, however, have obtained some knowledge concerning the types of conduct which the G.M.C. considered unacceptable. These included cases where practitioners had been convicted of criminal offenses such as theft, fraud, forgery, perjury, abortion, indecent assault, attempted sodomy, and arson. There were also cases in which practitioners had been found guilty of "infamous conduct in a professional respect." These involved such matters as covering unqualified assistants, committing adultery with patients, publishing indecent work, improperly disclosing confidential details of a patient's condition, improperly using qualifications, and obtaining registration by fraudulent means.

As a way of alerting practitioners to the extent and nature of unprofessional conduct, these cases were not altogether satisfactory for a number of reasons. First, only a limited range of issues presented themselves for adjudication by the Council and clearly not every type of misconduct was dealt with. For example, prior to 1883, no practitioners had been found guilty of "infamous conduct in a professional respect" for advertising, canvassing, or depreciation of colleagues, although some complaints regarding such matters had been received ([6], March 3, 1864, III, p. 398; February 25, 1868, VI, p. 306; July 5, 1869, VII, p. 46; July 6, 1869, VII, pp. 49–50; July 7, 1869, VII, p. 57; July 12, 1869, VII, p. 121; December 16, 1869, VII, p. 3 ; Dec. 15, 1870, VIII, p. 6). Indeed, on 27 November 1893 the Executive Committee of the Council was able to resolve that no rule at that time had been laid down by the G.M.C. against advertising and that advertising in itself was not to be regarded as "infamous conduct in a professional respect" ([6], November 27, 1893, XXX, p. 266).

Second, although some details of cases were reported in the G.M.C.'s *Minutes*, it was not until April 1864 that reporters were permitted entry to disciplinary proceedings, subject to a power to exclude them whenever the Council saw fit ([6], April 26, 1864, III, p. 16; see also, [10], 1864, I, p. 501). The information publicly available to reporters was then extensively reported in the medical press with, for example, the G.M.C.'s President, Dr. Paget, claiming that "the agency of the press, giving publicity to our debates and proceedings, has, I believe, more than doubled the powers of the Council" ([6], July 9, 1874, XI, pp. 12–3; see also, [10], 1874, II, p. 64). Prior to 1874, however, few practitioners were likely to be aware of the G.M.C.'s views regarding particular types of conduct.

THE WARNING NOTICES 1883 TO 1914

During the 1870s, the G.M.C. started to deal with cases in which practitioners were charged with employing and covering unqualified assistants, and by 1883 the Council had considered nine such cases. Of these cases, only one resulted in the charge being found proved although in that case the practitioner's name was not erased from the Register ([6], July 4, 1882, XIX, pp. 70–6).

On 21 April 1883, following the adoption of a report by the G.M.C.'s Committee on the Employment of Unqualified Assistants by Registered Practitioners, the Council made the following resolution:

That the Council record on its Minutes, for the information of those whom it may concern, that charges of gross misconduct in the employment of unqualified assistants, and charges of dishonest collusion with unqualified practitioners in respect of the signing of medical certificates required for the purposes of any law or lawful contract, are, if brought before the Council, regarded by the Council as charges of infamous conduct under the Medical Act. ([6], April 20, 1883, XX, p. 91).

This, then, was the first formal indication which the G.M.C. gave of the definition and scope of infamous conduct in a professional respect. On 20 November 1886, the Council resolved that "steps should be taken with a view of making public the above resolution" ([6], Nov. 20, 1886, XXIII, p. 152). This was subsequently carried out in July 1887 by inserting the above resolution twice in each of the leading medical journals ([6], July 25, 1887, XXIV, p. 402; [1], 1887, 2, p. 248; [10], 1887, II, p. 229; [16], 1887, 95, p. 111; [5], 1887, 33, p. 293).

The Council continued to deal with cases of covering although it was not until November 1888 that practitioners' names were first erased from the Register for this offense ([6], November 28, 1888, XXV, pp. 84–5; November 29, 1888, XXV, pp. 88–90; November 30, 1888, pp. 91–2). In each case, however, the practitioners' names were restored to the Register after twelve months ([6], November 28, 1889, XXVI, pp. 156–9, 161–3).

Throughout the 1880s and 1890s, the Council continued to refine and elaborate upon its resolutions regarding the employment of unqualified assistants ([6], May 22, 1888, XXV, pp. 37–8; May 26, 1893, XXX, p. 64; November 23, 1897, XXXIV, pp. 121–2). By the turn of the century a formal "Warning Notice" was issued to all newly registered medical practitioners and, after 1920, reproduced in the front of the bound volumes of the Medical Register until 1958 (when it was replaced by the Notes of the Disciplinary Committee, and then, after 1963 in the Blue Pamphlet).

By 1914, when the "Warning Notices" were revised and consolidated, they contained brief advice on questions of certification, employment of unqualified assistants, sale of poisons, association with unqualified persons, and advertising and canvassing. In addition, the Notices stressed that they did not refer to every possible type of professional misconduct and that circumstances could arise in which questions of professional misconduct fell outside the categories listed ([6], June 1, 1914, LI, p. 54).

The particular matters which were included in the "Warning Notice" arose directly out of disciplinary cases which had already been dealt with by the Council, and thus represented a distillation of the ethical principles that emerged from those cases ([11], pp. 149–51; [2], p. 419). Thus, it was clear that the G.M.C. was not a Parliament for making professional law ([12], p. 2), and that the "Warning Notice" was not a law or regulation made by the Council; "it was merely a condensed statement of the successive judgments of the court" ([12], p. 12; see also, [8], pp. 39–41; [17], pp. 9–13). This point was emphasized in the early 1900s when, following the issue of the Council's "Warning Notice" of 2 December 1901, with respect to the employment of unqualified pharmaceutical assistants, numerous practitioners objected to the G.M.C.'s trying cases before declaring the offense illegal ([6], Dec. 2, 1901, XXXVIII, pp. 129–30). In 1903, the Council received a six page petition, signed by 133 practitioners resident in Scotland, objecting to the way in which the Council proceeded in the setting and monitoring of ethical standards of professional conduct ([6], Jan. 6, 1902, XXXIX, p. 278; May 28, 1902, p. 282; June 3, 1902, p. 100; Nov. 23, 1903, XL, pp. 249–53).

The Executive Committee of the Council replied by issuing the following resolution:

> The Executive Committee desire to point out that the General Council have no power to legislate or to issue resolutions binding upon the profession and having absolute prohibitive effect. And in view of their judicial functions in particular cases of professional misconduct, it is not desirable to pass a resolution condemning any practice in general terms until a series of cases decided before them has so clearly demonstrated the prevalence of that practice as to call in the opinion of the Council, for a Warning Notice to the profession ([6], Nov. 23, 1908, XLV, pp. 224–5).

PROBLEMS IN THE DECLARATION OF STANDARDS

It now remains to consider whether or not the G.M.C.'s approach to the declaration of acceptable standards of professional conduct – the extraction of ethical principles from disciplinary cases and the issue of statements summarizing those principles after a series of cases has been disposed of – is the most efficient and effective way of proceeding.

1. Lapse of Time

The first difficulty which, arguably, arises is that there was often a considerable lapse of time between the appearance of the G.M.C.'s "Warning Notice" with respect to a given issue, and the initial hearing of disciplinary cases relating to that particular matter. Figure 1 depicts this problem diagramatically (see also [20]).

The vertical axis to this figure shows the present range of disciplinary matters which are contained in the Blue Pamphlet, while time is indicated on the abscissa. For each type of conduct three dates are provided: the date upon

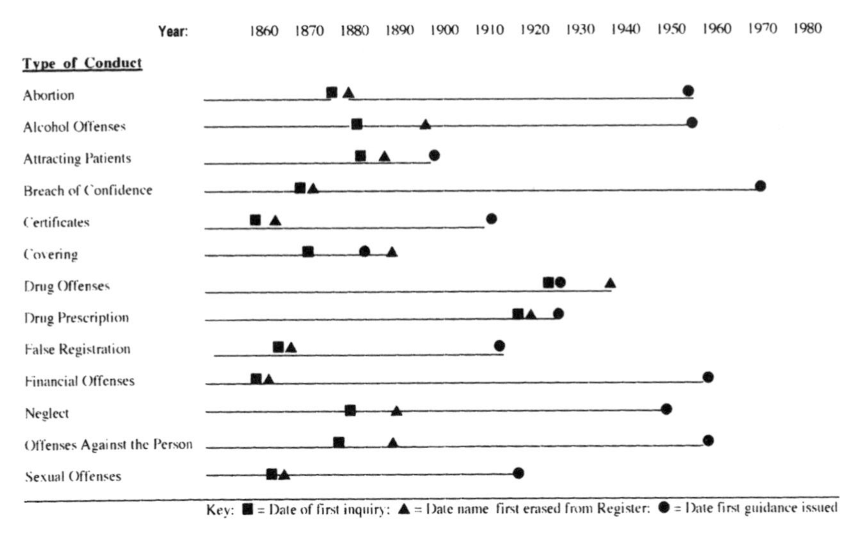

*Figure 1.*Time lapse between hearing of initial disciplinary case, erasure, and guidance.

which a disciplinary inquiry was first conducted (indicated by a ■); the date upon which a practitioner's name was first erased from the Register for that type of misconduct (indicated by a ▲); and the date upon which the G.M.C. first issued its "Warning Notice" or "Guidance" with respect to that type of misconduct (indicated by a ●).

For example, the first disciplinary inquiry involving a charge of "covering" was heard on 5 July 1871 ([6], IX, pp. 17–20); the first instance in which a practitioner's name was erased from the Register for covering occurred on 28 November 1888 ([6], XXV, pp. 84–5); while the G.M.C. first issued its "Warning Notice" with respect to covering on 20 April 1883 ([6], XX, p. 91). In all but two types of case, namely those relating to drug prescription and drug abuse, the Council had dealt with cases and erased names prior to issuing its "Warning Notices," with often many years transpiring in the interim period. A striking example relates to charges of "breach of confidence:" the first inquiry into which was held on 5 July 1869 ([6], VII, pp. 41–5), but the G.M.C. did not issue its formal guidance on breach of confidence until 24 November 1970 – a lapse of some 101 years ([6], CVII, p. 29). In the case of "drug offenses," however, a different situation was obtained: guidance was actually issued some six months *prior* to the first disciplinary inquiry. The dates and periods of time involved for each type of conduct are presented in Table 1.

Table 1. Dates of first inquiry, first erasure, and guidance by the GMC under the Medical Act of 1858, Section 29.*

Type of Conduct	First Inquiry	First Erasure	First Guidance	Years Between Inquiry & Guidance
Abortion	1878	1878	1958	80
Alcohol Offenses	1880	1893	1958	78
Attracting Patients	1884	1885	1899	15
Breach of Confidence	1869	1869	1970	101
Certification	1859	1861	1911	52
Covering	1871	1888	1883	12
Drug Offenses	1923	1936	1923	–.05
Drug Prescription	1917	1917	1923	6
False Certificates	1859	1861	1911	52
False Registration	1859	1859	1911	52
Financial Offenses	1861	1861	1958	97
Neglect of Patients	1879	1891	1951	73
Offenses Against the Person	1876	1888	1958	82
Sexual Offenses	1863	1863	1914	51

* Only cases involving convictions and improper conduct are included.

Although the G.M.C. has always stated that the matters dealt with in the "Warning Notice" were not to be taken as an exhaustive list, practitioners could, arguably, feel that they had been proceeded against before the conduct in question had been professionally proscribed.

2. Absence of Explanation

The second difficulty that arises relates to the fact that the G.M.C. has always maintained a strict policy of not elaborating upon, or explaining, the matters dealt with in "Warning Notices," because it believed that its judicial function was inconsistent with playing an advisory role. Accordingly, practitioners who contemplate embarking upon a course of action which may potentially bring them within the G.M.C.'s disciplinary remit, are unable to ascertain in advance whether or not the proposed course of action will, or will not, be viewed by the G.M.C. as "unprofessional." This problem continues to cause concern even today: for example, one practitioner, who was disciplined in 1983 for issuing prescriptions for controlled drugs otherwise than in the course of bona fide treatment, had requested guidance *prior* to embarking upon the treatment in question, only to be informed that "the Council has hitherto issued no specific guidance on that subject" ([3], p. 113).

Even where disciplinary proceedings have been commenced, practitioners may feel that the G.M.C. has been disingenuous in identifying the precise objections to the practitioner's conduct. For example, one practitioner wrote to the G.M.C.'s solicitor on 26 June 1895 seeking an explanation as to why he had been charged with publishing and circulating a book entitled "Electro-Homoeopathic Medicine:"

I am left entirely in the dark as to what kind of objections are found by the General Medical Council in these passages, and how I am to answer them ([6], July 22, 1895, XXXII, p. 194).

3. *Dissemination of Guidance*

In order for the G.M.C. to fulfill its role of setting and maintaining professional standards, it is essential that its proceedings and debates be widely disseminated and brought to the attention of all registered practitioners. As already mentioned, the G.M.C.'s debates prior to 1864 were closed to members of the public, including registered practitioners other than members of the Council. It was only after a concerted campaign by one G.M.C. member, Dr. Andrew Wood, in the 1860s, that press reporters were finally permitted to report certain parts of the G.M.C.'s debates ([10], I, 1864, p. 501). The early reports of proceedings in the medical and lay press were quite extensive: for example, in 1879 a report of the Executive Committee on the constitution and working of the G.M.C. was able to conclude that:

the admission of reporters has made the profession fully cognizant of the proceedings and debates, which have for the most part been published at length in the medical journals ([6], March 18, 1879, XVI, p. 24).

Unfortunately, over the 137 year history of the G.M.C.'s existence, there has been a steady decline in the extent to which information is reported with respect to the G.M.C.'s activities. (Some of the reasons for the *British Medical Journal*'s reticence are explored by Dr. Peter Bartrip in Chapter Nine of this volume). If practitioners are expected to refrain from the commission of professional misconduct, it is essential that they be provided with reasonably full and adequate reports of instances in which their professional colleagues have fallen foul of the G.M.C.'s professional conduct jurisdiction.

4. *Interpretation and Application of Guidance*

A number of serious difficulties exist in relation to the extraction of accurate, consistent, and workable ethical principles and rules of practice from decided cases, and also with respect to the interpretation of the G.M.C.'s guidance.

A. Absence of Reasoned Decisions

Section XXIX of the 1858 Medical Act originally only gave the General Council power to direct the Registrar to erase the name of a practitioner from the Register, and the early Standing Orders which regulated disciplinary procedure merely specified the form of resolutions to be voted upon with respect to whether or not a conviction was proved, whether or not the offense amounted to infamous conduct in a professional respect, and whether or not a direction for erasure should be given. As such, there was little scope given for the President to make comments or give reasons for decisions arrived at, although by 1932 the Standing Orders provided for the announcement of decisions by the President together with "such terms of reprimand, admonishment or otherwise as the Council shall approve" ([6], 1932, LXIX, Appendix XI, p. 343). In the case of a Dr. Theobald, whose name had been erased in 1894 for publishing a book on homeopathy, the practitioner had requested a rehearing on the grounds, *inter alia*, that the Council had failed to give reasons for its decision ([6], Dec. 3, 1894, XXXI, pp. 159–60; July 22, 1895, XXXII, p. 195). The G.M.C.'s solicitor, Mr. Farrer, however, confirmed that the Council "does not give reasons for or explain its decisions" ([6], July 22, 1895, XXXII, p.195).

This practice seriously detracts from the ability of the accused practitioner, other medical practitioners, and members of the public to understand and interpret the G.M.C.'s ethical determinations. The failure to give reasoned decisions may well leave the profession in some doubt as to what behavior constitutes professional misconduct and what particular sanctions have been imposed. In addition, the giving of reasoned decisions would greatly strengthen the appeal process by enabling the appellate body to know precisely what facts have been found proved, what view G.M.C. members took regarding conflicting evidence, and what particular considerations were taken into account in mitigation or aggravation as affecting the choice of a particular sanction.

Some of the arguments that have been raised against the giving of reasoned decisions include the possibility that individual G.M.C. members would differ on the reasons given, it would be difficult to prepare a reasoned decision with which everyone agreed in sufficient time to be delivered at the conclusion of the hearing, and that the preparation of such a decision would entail drafting by the legal assessor with whose views everyone might not agree (see, [18], p. 106, para. 309). In addition, there is the fear that the introduction of reasoned decisions would open the floodgates to all kinds of excessively legalistic and technical arguments, and would lead to decision-makers being obliged to have regard to prior decisions as precedents. As will be argued presently, the emergence of a jurisprudence based on prior decisions could be seen as a wholly worthwhile trend to be encouraged.

On balance, it appears that the arguments advanced in support of the obligation to give reasoned decisions outweigh those against and it is clear that the G.M.C.'s function of declaring ethical principles of good professional conduct would be enhanced if reasoned decisions were given in disciplinary cases.

B. Ad hoc Cases

A further difficulty with using disciplinary cases to declare principles of professional conduct is that the cases which result in public hearings tend to be ad hoc, disparate, and relate to their own peculiar factual circumstances. From the earliest times the G.M.C. has emphasized that it will not act as a police force for the profession in respect of discovering instances of misconduct, but rather, cases are brought to the attention of the Council by independent public authorities, such as the courts, or by individual complainants such as colleagues or patients ([6], May 16, 1862, II, pp. 94–5; June 27, 1882, XIX, p. 17; May 22, 1888, XXV, p. 18; 1926, LXIII, Appendix IV, p. 268; June 3, 1969, CVI, p. 9; June 2, 1970, CVII, p. 7; 1973, CX, Appendix III, pp. 186–7). Many factors are responsible for prompting individuals or bodies to report cases to the Council although generally dictates of fashion and topicality are important and these often follow closely new legislative reforms and contemporary social events or medical developments. For example, the occurrence of wars and changes in immigration patterns, and medical matters such as the introduction of vaccination, employment of medically unqualified assistants, the introduction of midwives, developments in cosmetic surgery, and problems of alcohol and drug abuse have all been associated with the incidence of cases of misconduct being reported (see, [20], pp. 99–100).

In addition, once cases are notified to the Council, a filtering process takes place in which the vast majority of complaints are excluded as being unsuitable for public hearing or as not raising questions of misconduct within the jurisdiction of the Council. Accordingly, only a highly limited and selective range of issues are adjudicated publicly and it is only these which generally form the basis of the Council's ethical guidance.

C. Absence of Precedent

The third problem, already adverted to, is that the Council has almost never relied upon the doctrine of precedent in deciding cases but rather considers each new case in isolation and without reference to cases of a similar nature which have arisen in the past. The reasons for this relate primarily to the absence of detailed reasons for decisions being given in cases, the fact that decisions are given *extempore*, and the fact that decision-makers change frequently. Thus, as the *Lancet* observed in discussing the case of Dr. Tarnesby in 1969:

> An evolving system of case law such as that from which our common law continues to develop is not to be extracted from the records of the Committee; and a lawyer, seeking to establish from these a coherent pattern, is likely to withdraw baffled ([10], 1969, II, p. 305).

As a means of creating an ongoing body of ethical principles and practical guidelines that have emerged from individual cases, the system that has evolved is not altogether satisfactory in this regard. Generally, then, for these reasons, it is, arguably, both inefficient and ineffective to attempt to declare ethical principles and guidelines for professional conduct through the analysis of prior disciplinary decisions of the Council and its Committees.

5. Unfair to Individual Practitioners

Finally, it seems to be unfair to require individual practitioners to undergo emotionally and financially burdensome disciplinary proceedings in order for general ethical and professional principles to be declared that will be of benefit and use to the whole professional community. Rather, disciplinary proceedings should only be used once the principles and guidelines have been established and alleged breaches of them identified.

LESSONS FROM THE PAST

In many respects the G.M.C.'s professional conduct jurisdiction is little different today from the penal jurisdiction of the nineteenth century. However, one important difference as already mentioned, is that the Council now has a legislative power to provide advice to members of the profession on standards of professional conduct and medical ethics. In addition, the G.M.C. is attempting to improve relations between itself, the profession, and the public by, for example, the appointment of a Press Officer and the provision of explanatory notes to members of the public in hearings.

Rather than perpetuate the manner in which ethical principles were extracted and declared in the nineteenth century, it would, arguably, be preferable for the Council's Standards Committee to declare principles of professional conduct and ethics in novel areas of medical practice in advance of the Council embarking upon disciplinary inquiries. In addition, there seem to be good reasons for the Council actively to offer advice to practitioners with respect to the scope and meaning of its guidance and to advise practitioners with respect to the acceptability or otherwise of given conduct. Finally, it seems that those involved in decision-making within the G.M.C. should show a greater willingness to make use of precedent and to give reasoned decisions in cases.

NOTE

[1] We are grateful to the copyright holder, the Wellcome Trust, for granting us permission to print this chapter; an earlier version was published in *Medical History* under the title "The Development of Ethical Guidance for Medical Practitioners by the General Medical Council" (1993, 37: 56–67).

BIBLIOGRAPHY

1. *British Medical Journal.*
2. Carr-Saunders, A. M. and Wilson, P. A.: 1933, *The Professions*, Clarendon, Oxford.
3. Daly, A.: 1990, *A Doctor's Story*, Macmillan London Ltd., London.
4. Draper, M.: 1980, "Elections to the G.M.C.," in General Medical Council, *Annual Report 1979*, p. 8.
5. *Edinburgh Medical Journal.*
6. G.M.C *Minutes.*
7. General Medical Council: January 1993, *Professional Conduct and Discipline: Fitness to Practise*, London.
8. Happel, J. S.: 1985, "Advice on good practice from the Standards Committee", *Journal of Medical Ethics* 11, 39–4.
9. House of Lords: 1950 Parliamentary Debates, 166, 18 April 1950.
10. *Lancet.*
11. Little, E.G.: 1926 "The General Medical Council", *The Nineteenth Century and After*, DLXXXVIII, 149, 151.
12. MacAlister, D.: 1906, *Introductory Address on the General Medical Council: Its Powers and Its Work*, Manchester University Press, Manchester.
13. Medical Act 1858 (21 & 22 Vict. c. 90).
14. Medical Act 1950 (14 Geo. 6 c. 29).
15. Medical Act 1983 (c. 54).
16. *Medical Press and Circular.*
17. Pyke-Lees, W.: 1958, *Centenary of the General Medical Council, 1858–1958.*
18. *Report of the Committee of Inquiry into the Regulation of the Medical Profession* (Chairman: Dr. A. W. Merrison) 1975, Cmnd. 6018, H.M.S.O., London.

19. Shaw, G. B.: 1931, *Doctors' Delusions*, Constable & Co. Ltd., London, 47–8.
20. Smith, R. G.. 1994, *Medical Discipline: The Professional Conduct Jurisdiction of the General Medical Council, 1858–1990*, Clarendon, Oxford.
21. *The Queen, On the Prosecution of Richard Organ* v *General Council of Medical Education and Registration of the United Kingdom*: 1861, 30 L. J. (Q. B.) 201.

NOTES ON CONTRIBUTORS

Robert Baker, Professor and Chair of the Department of Philosophy, Union College, Schenectady, New York.

John Bell (1796–1872), Professor of Medicine at the University of Pennsylvania, a member of the American Philosophical Society and of the College of Physicians of Philadelphia, and editor of the *Eclectic Journal of Medicine*.

Peter Bartrip, Senior Lecturer in History at Nene College, Northhampton, England.

Tom L. Beauchamp, Professor of Philosophy and Senior Scholar at the Joseph and Rose Kennedy Institute of Ethics, Georgetown University, Washington, D.C.

Chester R. Burns, James Wade Rockwell Professor of Medical History at the Institute for the Medical Humanities, University of Texas Medical Branch at Galveston.

M. Anne Crowther, Professor in the Department of Economic and Social History, University of Glasgow, Scotland.

Isaac Hays (1796–1879), physician to the Philadelphia Dispensary, first President of the Opthalmological Society of Philadelphia, first Treasurer of the American Medical Association, and founding editor of the *American Journal of the Medical Sciences*.

Stanley Joel Reiser, Director of the Program on the Humanities and Technology in Health Care at the University of Texas Health Care Science Center in Houston.

Russell Smith, Barrister and Solicitor of the Supreme Court of Victoria, a Solicitor of the Supreme Court of England and Wales, and a Lecturer in

R. Baker (ed.), The Codification of Medical Morality, 219–220.
© 1995 *Kluwer Academic Publishers. Printed in the Netherlands.*

Criminology in the Criminology Department of the University of Melbourne, Victoria.

Jukes Styrap (1815–1899), Member of the Royal College of Surgeons, a Licentiate of the Society of Apothecaries, a licentiate of the Royal College of Physicians of Ireland, and a founder of the Salopian Medico-Ethical Society.

Robert M. Veatch, Professor and Director of the Joseph and Rose Kennedy Institute of Ethics, Georgetown University, Washington, D.C.

INDEX

Philosophy and Medicine

1. H. Tristram Engelhardt, Jr. and S.F. Spicker (eds.): *Evaluation and Explanation in the Biomedical Sciences.* 1975 ISBN 90-277-0553-4
2. S.F. Spicker and H. Tristram Engelhardt, Jr. (eds.): *Philosophical Dimensions of the Neuro-Medical Sciences.* 1976 ISBN 90-277-0672-7
3. S.F. Spicker and H. Tristram Engelhardt, Jr. (eds.): *Philosophical Medical Ethics: Its Nature and Significance.* 1977 ISBN 90-277-0772-3
4. H. Tristram Engelhardt, Jr. and S.F. Spicker (eds.): *Mental Health: Philosophical Perspectives.* 1978 ISBN 90-277-0828-2
5. B.A. Brody and H. Tristram Engelhardt, Jr. (eds.): *Mental Illness.* Law and Public Policy. 1980 ISBN 90-277-1057-0
6. H. Tristram Engelhardt, Jr., S.F. Spicker and B. Towers (eds.): *Clinical Judgment: A Critical Appraisal.* 1979 ISBN 90-277-0952-1
7. S.F. Spicker (ed.): *Organism, Medicine, and Metaphysics.* Essays in Honor of Hans Jonas on His 75th Birthday. 1978 ISBN 90-277-0823-1
8. E.E. Shelp (ed.): *Justice and Health Care.* 1981 ISBN 90-277-1207-7; Pb 90-277-1251-4
9. S.F. Spicker, J.M. Healey, Jr. and H. Tristram Engelhardt, Jr. (eds.): *The Law-Medicine Relation: A Philosophical Exploration.* 1981 ISBN 90-277-1217-4
10. W.B. Bondeson, H. Tristram Engelhardt, Jr., S.F. Spicker and J.M. White, Jr. (eds.): *New Knowledge in the Biomedical Sciences.* Some Moral Implications of Its Acquisition, Possession, and Use. 1982 ISBN 90-277-1319-7
11. E.E. Shelp (ed.): *Beneficence and Health Care.* 1982 ISBN 90-277-1377-4
12. G.J. Agich (ed.): *Responsibility in Health Care.* 1982 ISBN 90-277-1417-7
13. W.B. Bondeson, H. Tristram Engelhardt, Jr., S.F. Spicker and D.H. Winship: *Abortion and the Status of the Fetus.* 2nd printing, 1984 ISBN 90-277-1493-2
14. E.E. Shelp (ed.): *The Clinical Encounter.* The Moral Fabric of the Patient-Physician Relationship. 1983 ISBN 90-277-1593-9
15. L. Kopelman and J.C. Moskop (eds.): *Ethics and Mental Retardation.* 1984 ISBN 90-277-1630-7
16. L. Nordenfelt and B.I.B. Lindahl (eds.): *Health, Disease, and Causal Explanations in Medicine.* 1984 ISBN 90-277-1660-9
17. E.E. Shelp (ed.): *Virtue and Medicine.* Explorations in the Character of Medicine. 1985 ISBN 90-277-1808-3
18. P. Carrick: *Medical Ethics in Antiquity.* Philosophical Perspectives on Abortion and Euthanasia. 1985 ISBN 90-277-1825-3; Pb 90-277-1915-2
19. J.C. Moskop and L. Kopelman (eds.): *Ethics and Critical Care Medicine.* 1985 ISBN 90-277-1820-2
20. E.E. Shelp (ed.): *Theology and Bioethics.* Exploring the Foundations and Frontiers. 1985 ISBN 90-277-1857-1
21. G.J. Agich and C.E. Begley (eds.): *The Price of Health.* 1986 ISBN 90-277-2285-4
22. E.E. Shelp (ed.): *Sexuality and Medicine.*
    Vol. I: Conceptual Roots. 1987 ISBN 90-277-2290-0; Pb 90-277-2386-9

23. E.E. Shelp (ed.): *Sexuality and Medicine.* Vol. II: Ethical Viewpoints in Transition. 1987 ISBN 1-55608-013-1; Pb 1-55608-016-6
24. R.C. McMillan, H. Tristram Engelhardt, Jr., and S.F. Spicker (eds.): *Euthanasia and the Newborn.* Conflicts Regarding Saving Lives. 1987 ISBN 90-277-2299-4; Pb 1-55608-039-5
25. S.F. Spicker, S.R. Ingman and I.R. Lawson (eds.): *Ethical Dimensions of Geriatric Care.* Value Conflicts for the 21th Century. 1987 ISBN 1-55608-027-1
26. L. Nordenfelt: *On the Nature of Health.* An Action-Theoretic Approach. 2nd, rev. ed. 1995 ISBN Hb 0-7923-3369-1; Pb 0-7923-3470-1
27. S.F. Spicker, W.B. Bondeson and H. Tristram Engelhardt, Jr. (eds.): *The Contraceptive Ethos.* Reproductive Rights and Responsibilities. 1987 ISBN 1-55608-035-2
28. S.F. Spicker, I. Alon, A. de Vries and H. Tristram Engelhardt, Jr. (eds.): *The Use of Human Beings in Research.* With Special Reference to Clinical Trials. 1988 ISBN 1-55608-043-3
29. N.M.P. King, L.R. Churchill and A.W. Cross (eds.): *The Physician as Captain of the Ship.* A Critical Reappraisal. 1988 ISBN 1-55608-044-1
30. H.-M. Sass and R.U. Massey (eds.): *Health Care Systems.* Moral Conflicts in European and American Public Policy. 1988 ISBN 1-55608-045-X
31. R.M. Zaner (ed.): *Death: Beyond Whole-Brain Criteria.* 1988 ISBN 1-55608-053-0
32. B.A. Brody (ed.): *Moral Theory and Moral Judgments in Medical Ethics.* 1988 ISBN 1-55608-060-3
33. L.M. Kopelman and J.C. Moskop (eds.): *Children and Health Care.* Moral and Social Issues. 1989 ISBN 1-55608-078-6
34. E.D. Pellegrino, J.P. Langan and J. Collins Harvey (eds.): *Catholic Perspectives on Medical Morals.* Foundational Issues. 1989 ISBN 1-55608-083-2
35. B.A. Brody (ed.): *Suicide and Euthanasia.* Historical and Contemporary Themes. 1989 ISBN 0-7923-0106-4
36. H.A.M.J. ten Have, G.K. Kimsma and S.F. Spicker (eds.): *The Growth of Medical Knowledge.* 1990 ISBN 0-7923-0736-4
37. I. Löwy (ed.): *The Polish School of Philosophy of Medicine.* From Tytus Chałubiński (1820–1889) to Ludwik Fleck (1896–1961). 1990 ISBN 0-7923-0958-8
38. T.J. Bole III and W.B. Bondeson: *Rights to Health Care.* 1991 ISBN 0-7923-1137-X
39. M.A.G. Cutter and E.E. Shelp (eds.): *Competency.* A Study of Informal Competency Determinations in Primary Care. 1991 ISBN 0-7923-1304-6
40. J.L. Peset and D. Gracia (eds.): *The Ethics of Diagnosis.* 1992 ISBN 0-7923-1544-8

Philosophy and Medicine

41. K.W. Wildes, S.J., F. Abel, S.J. and J.C. Harvey (eds.): *Birth, Suffering, and Death.* Catholic Perspectives at the Edges of Life. 1992 [CSiB-1]
ISBN 0-7923-1547-2; Pb 0-7923-2545-1
42. S.K. Toombs: *The Meaning of Illness.* A Phenomenological Account of the Different Perspectives of Physician and Patient. 1992
ISBN 0-7923-1570-7; Pb 0-7923-2443-9
43. D. Leder (ed.): *The Body in Medical Thought and Practice.* 1992
ISBN 0-7923-1657-6
44. C. Delkeskamp-Hayes and M.A.G. Cutter (eds.): *Science, Technology, and the Art of Medicine.* European-American Dialogues. 1993 ISBN 0-7923-1869-2
45. R. Baker, D. Porter and R. Porter (eds.): *The Codification of Medical Morality.* Historical and Philosophical Studies of the Formalization of Western Medical Morality in the Eighteenth and Nineteenth Centuries, Volume One: Medical Ethics and Etiquette in the Eighteenth Century. 1993 ISBN 0-7923-1921-4
46. K. Bayertz (ed.): *The Concept of Moral Consensus.* The Case of Technological Interventions in Human Reproduction. 1994 ISBN 0-7923-2615-6
47. L. Nordenfelt (ed.): *Concepts and Measurement of Quality of Life in Health Care.* 1994 [ESiP-1] ISBN 0-7923-2824-8
48. R. Baker and M.A. Strosberg (eds.) with the assistance of J. Bynum: *Legislating Medical Ethics.* A Study of the New York State Do-Not-Resuscitate Law. 1995 ISBN 0-7923-2995-3
49. R. Baker (ed.): *The Codification of Medical Morality.* Historical and Philosophical Studies of the Formalization of Western Morality in the Eighteenth and Nineteenth Centuries, Volume Two: Anglo-American Medical Ethics and Medical Jurisprudence in the Nineteenth Century. 1995
ISBN 0-7923-3528-7; Pb 0-7923-3529-5
50. R. Carson and C. Burns (eds.): *The Philosophy of Medicine and Bioethics.* Reprospective and Critical Appraisal. 1995 (forthcoming) ISBN 0-7923-3545-7
51. K.W. Wildes, S.J. (ed.): *Critical Choices and Critical Care.* Catholic Perspectives on Allocating Resources in Intensive Care Medicine. 1995 [CSiB-2]
ISBN 0-7923-3382-9

KLUWER ACADEMIC PUBLISHERS – DORDRECHT / BOSTON / LONDON